COSMOLOGIES

COSMOLOGIES

The proceedings of the seventh annual conference of the Sophia Centre for the Study of Cosmology in Culture, University of Wales, Trinity Saint David, 6-7 June 2009.

Edited by Nicholas Campion

SOPHIA CENTRE PRESS

First published 2010.

Sophia Centre Press
University of Wales, Trinity Saint David,
Ceredigion, Wales SA48 7ED, United Kingdom.
www.sophiacentrepress.com

ISBN 978-1-907767-00-5

British Library Cataloguing in Publication Data.
A catalogue card for this book is available from the British Library.

Printed in the UK by CPI Anthony Rowe, Chippenham.

Contents

Images

Page 16, Figure 1.1: The Avebury Cove today. Photograph reproduced with kind permission from Lionel Sims.

Page 16, Figure 1.2: Arial diagram of the Avebury Cove within the Avebury henge and circle, with North's interpretation of lunar-solar alignments. From John North, *Stonehenge: Neolithic Man and the Cosmos* (London: Harper Collins, 1996), p. 274. Reproduced with kind permission from HarperCollins Publishers, London.

Page 17, Figure 2.1: Arial diagram of the Longstones Cove (L16, L15, L14 'Adam', L11) in the Beckhampton Avenue. From M. Gillings, J. Pollard, D. Wheatley and R. Peterson, *Landscape of the Megaliths: Excavation and Fieldwork on the Avebury Monuments, 1997-2003* (Oxford: Oxbow, 2008), p. 63. Reproduced with kind permission from Oxbow Books, Oxford.

Page 17, Figure 2.2: The Adam Stone, L14. From M. Gillings, J. Pollard, D. Wheatley and R. Peterson, *Landscape of the Megaliths: Excavation and Fieldwork on the Avebury Monuments, 1997-2003* (Oxford: Oxbow, 2008), p. 86. Reproduced with kind permission from Oxbow Books, Oxford.

Page 18, Figure 3: Stukeley's panorama of the West Kennet Avenue, showing the Cove at the 'apex' to the Avenue (centre left). From P.J. Ucko, M. Hunter, A.J. Clark and A. David, *Avebury Reconsidered: from the 1660s to the 1990s* (London: Unwin Hyman, 1991), p. 191. Reproduced with kind permission from Taylor & Francis Books, UK.

Page 27, Figure 4: Alignments at the Longstones Cove. From M. Gillings, J. Pollard, D. Wheatley and R. Peterson, *Landscape of the Megaliths: Excavation and Fieldwork on the Avebury Monuments, 1997-2003* (Oxford: Oxbow, 2008), p. 63. Reproduced with kind permission from Oxbow Books, Oxford.

Page 29, Figure 5: The Avebury complex. Image adapted from Pete Glastonbury, *Avebury Panoramic Tour,* CD ROM (2001). Reproduced with kind permission from Pete Glastonbury.

Acknowledgements

With grateful thanks to the Sophia Trust for its generous funding of the Sophia Centre. Also with thanks to Alice Ekrek for her editorial assistance in seeing this book from conception to publication.

Introduction: Cosmologies

Nicholas Campion

This volume contains the proceedings of the seventh annual conference of the Sophia Centre, formerly located at Bath Spa University and now, since 2007, at the University of Wales, Trinity Saint David. It is the third to be published, the first two being *Astrology and the Academy* and *Sky and Psyche*.[1]

The purpose of the Sophia Centre, as is evident in its full title, is the 'study of cosmology in culture', that is, the manner in which human beings relate their cultures to their notions of the nature, order, function or meaning of the cosmos. Traditionally, everything is included in the cosmos, including humanity, which is itself, therefore, an integral part of the 'everything'. In this sense, a study of one's self is a study of the cosmos, arguably a controversial notion in the context of modern scientific cosmology, but not to anyone schooled in certain of the world's philosophical conventions including, in the west, the Platonic. As with the Traditional Cosmology Society, whose name makes their interests clear, cosmology deals with the way in which humans orient their culture, their buildings, practices, beliefs and rituals, to the cosmos. The Society's remit is suitably broad, as described on its website:

> The society is concerned with exploring myth, religion and cosmology across cultural and disciplinary boundaries and with increasing understanding of world views in the past and present.[2]

In the overall context of research priorities at the University of Wales, Trinity Saint David, cosmology is situated within a wider project of research into landscape and the environment. In this context the Sophia Centre has contributions to make which are unique within the university sector, both of which are located in traditional and pre-modern understandings of cosmology. Firstly, the Centre extends the notion of environment to include what we might call the 'skyscape', the powerful but, in the modern

---

1 Nicholas Campion, Patrick Curry and Michael York, eds., *Astrology and the Academy, papers from the inaugural conference of the Sophia Centre, Bath Spa University College, 13-14 June 2003* (Bristol: Cinnabar Books, 2004); Nicholas Campion and Patrick Curry, eds., *Sky and Psyche: the Relationship between Cosmos and Consciousness* (Edinburgh: Floris Books, 2006).

2 *The Traditional Cosmology Society* website, http://www.tradcos.co.uk/ [accessed 4 February 2010].

artificially-lit world, often-ignored, patterns of the sun, moon, planets and stars; no study of the traditional environment can satisfactorily neglect this central feature of the pre-modern worldview. Secondly, the Centre's work recognises that in pre-modern psychology the environment, especially the heavens, or 'skyscape', was interiorised, both through classical thought and indigenous philosophies, so that there was a constant, mutually interdependent and dialectical relationship between the external environment and the individual, inner world. There is little, in this understanding of the term, which is not cosmological. Hans Jonas summed up one similarly broad version of the classical approach well made in his study of the Gnostics, whose cosmology he described as follows:

> By a long tradition this term [cosmos] had to the Greek mind become invested with the highest religious dignity. The very word by its literal meaning expresses a positive evaluation of the object — any object — to which it is accorded as a descriptive term. For *cosmos* means 'order' in general, whether of the world or a household, of a commonwealth, of a life: it is a term of praise and even admiration.[3]

In the modern sense, though, cosmos, as a 'meaning-ful' place in which humanity is an active participant, has generally been replaced by universe as a 'meaning-less' arena in which physical laws work themselves out. I place the term 'meaning-less' in parentheses because modern cosmology clearly carries meaning for many people, but no longer as a series of omens, instructions to perform ritual acts, paths to salvation or the unfolding of divine plans. In common usage, whereas cosmos in the traditional sense included inner space, the human psyche, now the universe, is confined to outer space, for which it is sometimes a synonym.

Cosmology is, as Norris Hetherington, something of an authority in the field, put it,

> the science, theory or study of the universe as an orderly system, and of the laws that govern it; in particular, a branch of astronomy that deals with the structure and evolution of the universe.[4]

This is, of course, a modern view, emphasising the *logos* of cosmology as a study, as the rational, scientific and technological investigation of the cosmos. In a different context the *logos* might be the word, which is how it is translated in the famous opening passage of John's Gospel, suggesting that the cosmos is an entity which speaks to humanity.[5]

3 Hans Jonas, *The Gnostic Religion: The Message of the Alien God and the Beginnings of Christianity*, second edition (Boston: Beacon Press, 1963), p. 241.
4 Norris D. Hetherington, *The Encyclopaedia of Cosmology: Historical, Philosophical and Scientific Foundations of Modern Cosmology* (New York, 1993), p. 116.

The papers in this volume deal, from a variety of perspectives, with different approaches to cosmology, a plurality which is summed up in the title, 'Cosmologies'. If the approaches we publish here are linked by a common thread, it's that they are all historical or anthropological, but they move from Megalithic monuments to the Dead Sea Scrolls to Cyberspace, while the research methods vary from textual analysis to participant observation.

There is one change to the published version of the conference proceedings (Lionel Sims was unable to submit his conference paper on 'Stonehenge Decoded: the Conflation of Winter Solstice Sunset with the Southern Minor Standstill Moonsets', so substituted the paper published here, 'Coves, Cosmology and Cultural Astronomy'), while three of the papers presented at the conference were not submitted for publication (Jenny Blain, 'Northern European Cosmologies of the Tree and the Well', Peter Forshaw, 'Astronomia Inferior et Superior: Some Medieval and Renaissance Instances of the Conjunction of Alchemy and Astrology' and Elizabeth Reichell, 'The Landscape in the Cosmoscape: Cosmology, Ethnoastronomy and Socio-environmental Sustainability Among the Tanimuka and Yukuna, N/W Amazon'), and Jarita Holbrook's ('Astronomy & Astrology in "the Sky in Our Lives" Survey') was submitted to *Culture and Cosmos*. All these will be missed.

The Sophia Centre is now in its eighth year of operation and its third year at Lampeter and is developing a variety of new initiatives, including specific research projects. All who are interested are invited to visit the Centre's website at: www.lamp.ac.uk/sophia.

Bibliography

Campion, Nicholas, Patrick Curry and Michael York, eds., *Astrology and the Academy, papers from the inaugural conference of the Sophia Centre, Bath Spa University College, 13-14 June 2003* (Bristol: Cinnabar Books, 2004).

Campion, Nicholas and Patrick Curry, eds., *Sky and Psyche: the Relationship between Cosmos and Consciousness* (Edinburgh: Floris Books, 2006).

Hetherington, Norris D., *The Encyclopaedia of Cosmology: Historical, Philosophical and Scientific Foundations of Modern Cosmology* (New York, 1993).

Jonas, Hans, *The Gnostic Religion: The Message of the Alien God and the Beginnings of Christianity*, second edition (Boston: Beacon Press, 1963).

The Traditional Cosmology Society website, http://www.tradcos.co.uk/ [accessed 4 February 2010].

Coves, Cosmology and Cultural Astronomy

Lionel Sims

Cultural astronomy has a key role to play in interpreting how every culture tries to make sense of 'life' through its 'cosmology'. This paper argues that this role can only be achieved once scholarship is able to transcend both the post-modern critique and the over-narrow definition of field method current within archaeoastronomy. By adopting a multi-disciplinary approach, or an American definition of anthropology, different methodologies can be brought to bear on any culture's cosmology which, when combined, achieve 'emergent' properties which exponentially reduce the possible number of testable interpretations. This methodology is demonstrated through a new interpretation of 'coves' which is consistent with a recent 'transformational template' of lunar-solar conflation.

Introduction

Every culture has a 'cosmology' — a theory that integrates the sum total of experience with the collectively represented origins and nature of all their known worlds. These worlds may be underworld(s)/this world/above world(s) in pre-state societies, and multi-verse or any combination of all in the west. Many cosmologies are also religions. At the Sophia Centre conference in Bath in June 2009, two seemingly unrelated contributions qualified what we might ever be able to say about 'cosmologies'. The theme of the conference was to explore whether anthropology and cultural astronomy could generate relevant definitions of and methodological approaches to the cosmologies of all cultures. Particular attention was paid to what future there might be for cultural astronomy in the light of the unresolved issue within anthropology of defining culture and the present restriction of the concept of cosmology to 'cold' or static small-scale or traditional societies. Within these debates a post-modernist contribution insisted that all theories of cosmology are just stories that tell us nothing about the world but a lot about the politics of the teller of the story,[1] and an archaeoastronomer argued that cultural astronomy is unable to interpret 'coves' since they are probably aligned on local landscape features and do not have the general relevance they might have if they were aligned on sun,

1 Panel contribution by Patrick Curry at the Sophia Centre 'Cosmologies' Conference, Bath, 6 June 2009.

moon or stars.[2] Therefore, at a conference dedicated to find a general paradigm for the future of cultural astronomy, one contribution denied the status of any truth claims, and another questioned the ability of cultural astronomy to interpret the horizon alignments of some of the largest structures of prehistory in the British Isles. Any discipline is judged by its ability to handle anxieties such as these, and this paper suggests that innovative interdisciplinary, inter-cultural and inter-institutional collaboration can answer such criticisms and provide a fruitful future for cultural astronomy.

Let us pose the problem of defining culture and cosmology as simply and as starkly as possible — are we 'lumpers' or 'splitters'? If we lump all the world's cosmologies together to find their common elements, then this might provide a way forward to a unified future for cultural astronomy. But the danger of amalgamation is that by abstracting all detail except that which is common will require jettisoning most of the ethnographic detail of each culture's cosmology — precisely the detail which carries cosmological meaning. All that would remain would be a few barren abstractions. Alternatively, if we retain all the distinctive detail of every culture's cosmology, then this may provide the evidence for a thick description of their various meanings, but at the expense of separating off each culture from each other as unique and idiosyncratic. With the first approach we gain commonality but lose meaning, with the second we never achieve common ground.

We can side-step this polarisation by acknowledging some recent discoveries of the life sciences — all of humanity shares a common and recent African origin. Fully modern cultural humans had evolved in sub-Saharan Africa by about 120,000 years ago, and the first group to successfully leave Africa did so about 80,000 years ago, carrying with them the genetic pool which accounts for all the out-of-Africa human variation today. All earlier hominids are not our ancestors, and the migrating wave of moderns rapidly displaced all earlier species of homo that left Africa, including Neanderthals in Europe. Therefore the biological differences between all of the world's people today are tiny and recent. More than that, the mit-DNA evidence suggests that our earliest ancestors were matrilocal — *husbands* moved into their *wives*' group. And one of the main pieces of evidence that these biological ancestors were culture-bearing is their systematic and sustained use of red-ochre for symbolic, not utilitarian, purposes. These findings are general for all early moderns in sub-Saharan Africa, and point towards the sharing of a common cultural 'package'.[3]

2 Statement made by Clive Ruggles in the film *Celebrating the Summer Solstice: The Pagan Experience*, directed by Darlene Villicana, shown at Sophia Centre 'Cosmologies' Conference, Bath, 6 June 2009 [hereafter Statement by Ruggles in Villicana, *Celebrating the Summer Solstice*].

3 F. D'Errico, 'The Invisible Frontier: a multiple species model for the origin of

These are observations — not theories. This evidence reduces the number of possible cultural origins scenarios. A theory rooted in patriarchal assumptions, for example, would find it difficult to absorb the evidence of matrilocality *plus* red-ochre use. A cognitive theory would find it difficult to explain how matrilineal/matrilocal coalitions were replaced later by patrilineal/patrilocal coalitions with no accompanying significant neural changes. A diffusionist theory would find it difficult to explain the very different dates and regional variation for the adoption of 'social complexity'. But we are a long way from sub-Saharan Africa before the last ice age; and even if this model of our recent origins is correct, all the world's cultures have since differentiated themselves from their common heritage. What might be the value of acknowledging a 'lumper' theory of a common culture and cosmology for our origins, and how might it be related to a 'splitter' theory that celebrates all subsequent diversity?

Transformational template

The present variety in the world's cosmologies cannot be explained, Mandelbrot-like, as ever more miniaturised or re-combined sets of the earliest cultural configuration. We cannot assume a chaos theory scenario of constant repetition of the same cultural rules at different levels of magnification. This approach works by formalistic regression, in which all later developments are reduced to their formal identity with the origins model, and abolishes the property of emergence in which new content suffuses old forms and qualitatively transforms social formations. But, equally, neither can we ignore the fact that all human cultures derive biologically and culturally from a common African heritage. For anthropology, and therefore cultural astronomy, to embrace all of the world's cosmologies in their specificity, and to demonstrate a common thread in our humanity which can be traced back to our common origins, then, just one option is left to transcend the lumper/splitter dichotomy — we need a *transformational template* that can both explain the generic culture from which all cosmologies have derived, and this template must be amenable and explanatory of all subsequent transformations to which it has been adapted and moulded. This paper will first consider whether this is a reasonable way to proceed, then critique the two challenges to this view made at the conference, provide a test of this critique through an

behavioral modernity', *Evolutionary Anthropology,* (2003), Vol. 12, pp. 188-202; Henshilwood & Marean, 'The Origin of Modern Human Behaviour: critique of the models and their test implications', *Current Anthropology,* (2003), Vol. 44(5), pp. 627-651; C. Knight, *Blood Relations* (New Haven: Yale University Press, 1995), [hereafter Knight, *Blood Relations*]; C. Knight, C. Power & I. Watts, 'The Human Symbolic Revolution: a Darwinian account', *Cambridge Archaeology,* (1995), Vol. 5, pp. 75-114 [hereafter Knight et al., 'Human Symbolic Revolution'].

interpretation of coves, and finally suggest a possible future for cultural astronomy.

Three examples demonstrate the use of a 'transformational template'. Levi-Strauss argued that nearly all of the variations in kinship and marriage systems of south-east Asia could be explained by sequences of small variations to an original patrilineal/patrilocal template.[4] This model has been criticised for ignoring many examples of women's power and control that cannot be explained assuming patriarchal beginnings. Nevertheless, the idea of a transformational template can be retained by positing instead an original matrilineal/matrilocal template. This allows the model to be reworked through positing the subsequent collapse of sororal solidarity and the emergence of male-led coalitions, so extending the transformational template backwards.[5] In a separate and later exercise, Levi-Strauss also showed that all the one-thousand or so Amerindian myths followed a common syntax of rules which bracketed dichotomous pairs of motifs into a single system.[6] It has been shown that the same grammar is exhibited in all Australian aboriginal myths and European fairy tales.[7] While the syntax (langue) is invariant, the political meaning (parole) is highly variable according to the socio-economic and political context of the myth-telling. While Levi-Strauss interpreted this invariant syntax as reflecting the neural wiring of the brain, he also showed that the substantive content of this structure followed the universal theme of a male matriarchy myth — that women's primordial rule had to be overthrown by men to guard against women's chaotic inability to ensure cosmic order. Levi-Strauss' alliance theory of cultural origins, in which groups of brothers traded their sisters, could only allow an extreme misogynist theory of human cultural origins. It could not explain matriliny, matrilocality, and many other aspects of ritual which indicated women's power and role in human cultural origins. Nor did Levi-Strauss suggest a convincing explanation of why cultural origins had to begin with the male oppression of women. If instead the evidence and theory for primordial matrilineal/matrilocal clans is accepted, then this contrary evidence can be accommodated within a model of a subsequent counter-revolution against women. Once this amendment is made to his template, it can then include this otherwise unexplained anomalous data.

A third example is lunar-solar conflation theory from cultural astronomy. Six regional groups of monuments of the late Neolithic/Early Bronze Age in the British Isles can be seen to be sharing the same 'astronomical' syntax of lunar-solar conflation that derives from an original

---

4 C. Levi-Strauss, *The Elementary Structures of Kinship.*

5 Knight, *Blood Relations.*

6 C. Levi-Strauss, *The Naked Man: Mythologiques, Volume 4*, trans. John and Doreen Weightman (Chicago: Chicago University Press, 1990).

7 C. Knight, *Menstruation and the Origins of Culture* (London: University College London [unpublished PhD thesis], 1987).

lunar template that is being confiscated by an emerging male-led solar cult. Instead of monthly dark moons being the trigger for matrilineal seclusion rituals, now binary solstice dark moon rituals staged within monumental architecture are monopolised by a male shaman/priestly cult.[8]

These findings and interpretations are examples of 'transformational templates' that connect both a cultural origins scenario and explain the cultural resource and structure for subsequent cosmological modification. This realist position is able to combine a cultural origins model with the evidence for subsequent cosmological diversification, and is open to being tested against evidence.

Post-modernism

Post-modernism denies the validity of any attempt to make a truth-claim that derives from a meta-narrative. The assumptions of post-modernism are two-fold — one is that any interpretive claim is organised solely by the teller's political agenda, and secondly it assumes the null hypothesis — as we cannot directly observe reality then there is no observable order in the world outside of this politically imposed narrative. Of course, post-modernism is correct in that everything we say or theorise is embedded in a political and cultural context. We do not directly observe reality — we construct it to make it intelligible. But that is not all we do. By a process of abstraction we decompose a complex whole into its constituent elements, study their properties in isolation, and then consider the emergent effects of their reconstitution. The recombination of interpreted elements admits only a very few arrangements, and these models of reality are the interpreted whole. We never see 'capitalism' directly, since this is an abstraction that some of us have constructed to make sense of the perceived order and connections of many other small observations. Whether we are right or not to impose this pattern called 'capitalism' is debatable and open to peer-review. But there is another arbiter of our truth claims — experience. Since societies and cultures vary and undergo change, then that very variation and change sifts out those elements and brings to the fore the key organising principles of a culture and cosmology, and this confirms or weakens whatever interpretations we made earlier. Therefore, variability and change are a resource to test our initial interpretations of any order or pattern we observe in the world.

8 L.D. Sims, 'The Solarization of the Moon: manipulated knowledge at sarsen Stonehenge', *Cambridge Archaeological Journal,* (2006), Vol. 16(2), pp. 191-207 [hereafter Sims, 'Solarization of the Moon']; L.D. Sims, 'What is a Lunar Standstill? Problems of accuracy and validity in the Thom paradigm', *Mediterranean Archaeology & Archaeometry,* (2007), Vol. 6(3), pp. 157-163 [hereafter Sims, 'Lunar Standstill']; L.D. Sims, 'Entering, and Returning from, the Underworld: reconstituting Silbury Hill by combining a quantified landscape phenomenology with archaeoastronomy', *Journal of the Royal Anthropological Institute,* (2009), Vol. 15(2), pp. 386-408 [hereafter Sims, 'The Underworld'].

To be consistent, the post-modernist critique must be self-referential. If anyone attempting to make a truth claim about the world is driven by a political agenda, then that must also be true for a post-modernist who makes that claim. By the same token, any proponent of post-modernism is also engaged in a political strategy by telling a story that inhibits another speaker from making any truth claim about the world. Therefore, the political agenda of a post-modernist is against anyone asserting that the world is ordered in a certain way. Since to deny that they also operate a meta-narrative is tautological, the choice for post-modernism is therefore either nihilism or self-abolition. And as post-modernists do try and say *this* something about the world it is inconsistent and disingenuous, since that assumes that since this is their preferred model then not all stories are equally weak or driven *solely* by a political interest. We all, including post-modernists, use different types of tests to discriminate between weak and strong arguments and theories. Some 'stories' are better than others, and by various logical and not-so-logical procedures we all evaluate what interpretations we are prepared to conditionally accept. And if a 'story' that we presently call a theory is eventually superseded by another later, when our grasp of reality has become stronger, then that does not mean that the earlier version of reality we used was always 'wrong'. Frequently, the new version has been able to include the previous version but now integrated into a more complete and rounded account that can simultaneously work on a higher level. For example, archaeology is going through a protracted and as yet unfinished critique and rejection of 'farming revolution theory'.[9] This suggests that culture and institutional order only begins in the Neolithic, and rises out of the surpluses of agriculture. It is variously patriarchal, cognitive and diffusionist, according to the author. This theory flies in the face of the anthropology of hunter-gatherers, which insists that culture is fully elaborated and modern humans fully evolved long before farming and the Neolithic. Yet a bronze technology is an 'advance' over a lithic technology, but according to our transformational template this 'advance' is embedded within a social and cultural decline in the democracy and egalitarianism of the matrilineal clans. Lunar-solar conflation theory therefore fits with a spiral rather than a unilineal model of historical change which can include both a 'matriarchal' egalitarian hunter-gatherer model and farming revolution theory, but only by transcending the limits of farming revolution theory.

The choice for all is whether to risk making a claim about the world, or whether to reside in a bunker of self-doubt honing our ability to discern our own and everybody else's political agenda. Clearly, if cultural astronomy and scholarship in general is to have any future, it lies with taking risks and in making testable interpretations of cosmologies. Let us see how confident

9 J. Thomas *Understanding the Neolithic* (London: Routledge, 1999), [hereafter Thomas, *Understanding the Neolithic*].

we can be in such an exercise by a multi-disciplinary study of coves, with particular reference to the role of cultural astronomy in such an exercise.

Archaeoastronomy

The cultural astronomy of prehistoric cultures is its sub-discipline – archaeoastronomy. Since the late 1970s a new generation of archaeoastronomers have critiqued and extended the work of Hawkins and Thom in the 1950s and 1960s, and re-set the discipline on firmer foundations.[10] In particular the professional umbrella organisations for archaeoastronomy – SEAC and ISAAC[11] – have encouraged field researchers to use statistical techniques on regional groups of monuments to test whether alignments found in any one monument are representative of the group. By dealing with aggregated sets of architecturally identical monuments from prehistory and using rigorous and pre-determined data selection and scaling procedures, this stage in archaeoastronomy has overcome much of the bias of previous research. However, the adoption of the statistical method was as much to do with overcoming a highly sceptical archaeological establishment and to establish a new academic discipline which they would accept had 'data'. Instead of the late twentieth century view within archaeology that 'astronomy' had no relevance to prehistoric cultures, it is now clear from a matured archaeoastronomy that most monuments in the Neolithic and Early Bronze Age (EBA) were intentionally designed with horizon alignments on cosmic bodies such as the sun, moon and stars.

But the statistical method has never been the only method available to archaeoastronomy. Some monuments are unique and highly complex, so much so that aspects of their detailed architecture will only allow an 'astronomical' interpretation. For example, the 'light box' above the entrance at Newgrange is accepted by all to be designed to let in the rays of winter solstice sunrise;[12] the grand trilithon double window at sarsen Stonehenge has only been explained by its double alignment of identity on winter solstice sunset and the southern minor standstill moonsets;[13] the Cove at the centre of the northern inner circle of the Avebury complex has

10 See J.W. Fountain & R.M. Sinclair, *Current Studies in Archaeoastronomy: conversations across time and space* (Durham: Carolina Academic Press, 2005); M. Hoskin, *Tombs, Temples and Their Orientations: a new perspective on mediterranean prehistory* (Bognor Regis: Ocarina, 2001); J. North, *Stonehenge: Neolithic Man and the Cosmos* (London: Harper Collins, 1996), [hereafter North, *Stonehenge*]; C. Ruggles, *Astronomy in Prehistoric Britain and Ireland* (London: Yale, 1999), [hereafter Ruggles, *Astronomy in Prehistoric Britain and Ireland*].

11 SEAC: http://www.archeoastronomy.org/; ISAAC: http://terpconnect.umd.edu/~tlaloc/archastro/

12 Ruggles, *Astronomy in Prehistoric Britain and Ireland*, p. 129.

13 North, *Stonehenge*; Sims, 'Solarization of the Moon'.

been shown to be a lunar-solar focussing device.[14] We can be confident that each of these interpretations are not committing 'the individualistic fallacy', since their architectural details are so unusual that no other hypothesis can displace these archaeoastronomical interpretations.

However, such has been the weight of archaeological disfavour that archaeoastronomy has stuck to accumulating aggregated data sets, and there has been very little development in the cultural interpretation of the alignments that have been found. A discipline that stands still waiting for others to accept it is a discipline in danger. As a sub-discipline of cultural astronomy, archaeoastronomy needs to widen its conceptual vocabulary to enhance its ability for cosmological interpretation. The puzzling case of coves allows us to test the challenge of post-modernism and the methods of archaeoastronomy for the possible futures of cultural astronomy.

Coves — testing cosmological concepts

Coves are found in monument complexes in the late Neolithic/EBA, and are tightly concentrated 'enclosures' of standing stones of three or four quadrangular orthostats in a rectangular or square plan. Very few coves are known of — it has been suggested that there were three at Avebury, and other coves were at Arbor Low, Stennes, Stanton Drew and Mount Pleasant in Dorset.[15] These structures include the largest stones ever moved in the prehistory of the British Isles — the back-stone of the Avebury Cove in the centre of the northern inner circle of the Avebury henge weighed about 100 tons.[16] Doubt mixed with wonderment surrounds any discussion of coves — Burl has referred to them as 'structural enigmas'.[17] Site excavation of coves confounds the archaeologists — the Longstones Cove is associated with a human 'burial' and thousands of pieces of worked flint, while the Avebury Cove is not associated with any deposition and seems to have been 'swept' clean. In a film shown at the Sophia Conference in Bath in June 2009, British archaeoastronomer Clive Ruggles referred to the diverse properties of 'coves', suggesting that such diversity amongst such a small sample defeated interpretation. This paper will concentrate on the three coves within the Avebury monument complex, and then look briefly at the remainder in the light of our analysis of the Avebury structures.

---

14 North, *Stonehenge*; also see the analysis which follows in this paper.

15 Burl, 'Coves: structural enigmas of the Neolithic', *Wiltshire Archaeological and Natural History Magazine*, (1988), Vol. 82, pp. 1-18 [hereafter Burl, 'Coves'].

16 M. Gillings, J. Pollard, D. Wheatley and R. Peterson, *Landscape of the Megaliths: Excavation and Fieldwork on the Avebury Monuments 1997-2003* (Oxford: Oxbow, 2008), [hereafter Gillings et al., *Landscape of the Megaliths*], pp. 62-90.

17 Burl, 'Coves'.

Figure 1.1: The Avebury Cove today.[18]

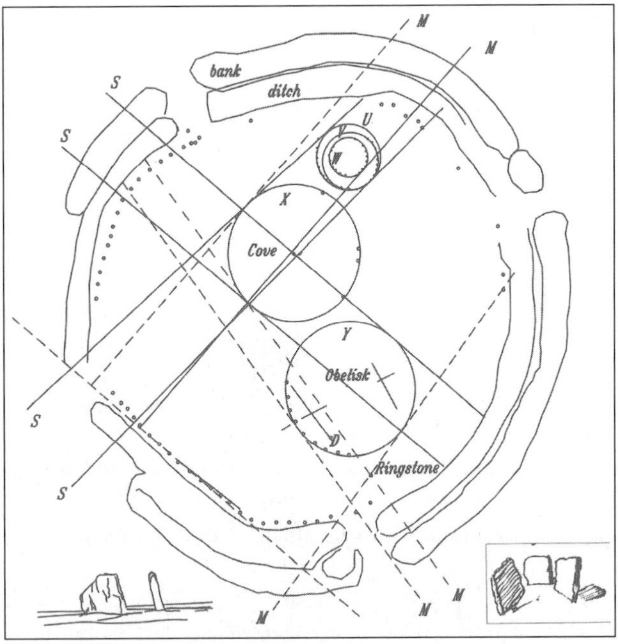

Figure 1.2: Arial diagram of the Avebury Cove within the Avebury henge and circle, with North's (1996) interpretation of lunar-solar alignments. Stukeley's sketches of the cove are inset.[19]

18 Author's photograph.
19 Image from North, *Stonehenge*, page 274.

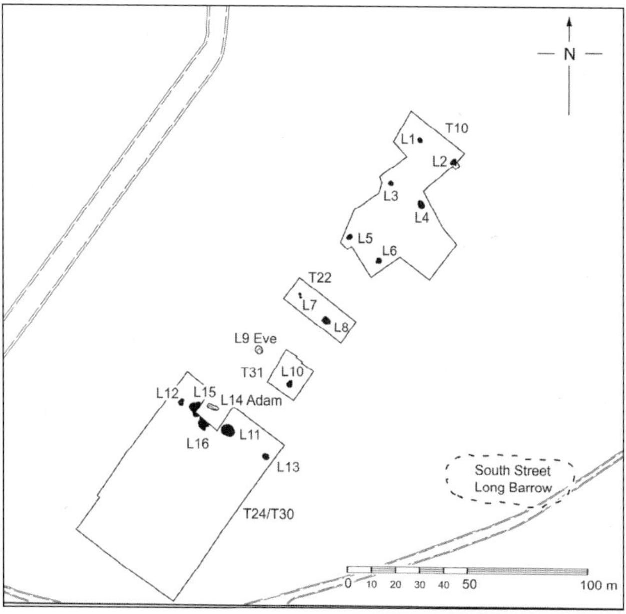

Figure 2.1: Arial diagram of the Longstones Cove (L16, L15, L14 'Adam', L11) in the Beckhampton Avenue.[20]

Figure 2.2: The Adam Stone, L14.[21]

20 Gillings et al., *Landscape of the Megaliths*, p. 63.
21 Gillings et al., *Landscape of the Megaliths*, p. 86.

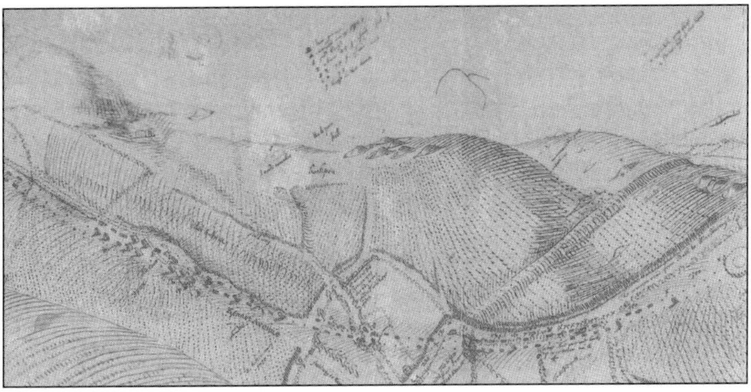

Figure 3: Stukeley's panorama of the West Kennet Avenue, showing the Cove at the 'apex' to the Avenue (centre left).[22]

The three coves thought to be at Avebury were the Avebury Cove at the centre of the northern inner circle of the Avebury great stone circle and henge, the Longstones Cove in the Beckhampton Avenue, and the Cove in the West Kennet Avenue (see Figs. 1, 2, 3). Of the first, there are just two stones still standing of what is thought to have been a three stone structure, within the much denuded northern inner circle and Avebury henge. The Longstones Cove has just one stone still remaining, where originally there had been four.[23] There is no remaining surface trace of the West Kennet Avenue cove, and we only know of it from the antiquarian testimony of William Stukeley.[24]

The first thing to note is that there is a pattern to the arrangement of stones, and it is only because of this pattern that we can come up with the category 'cove' — either four stones constitute a 'box', as at the Longstones Cove, or three stones form an open 'sentry box' arrangement, as at the centre of the northern inner circle and the West Kennet Avenue cove. They have been compared to the horseshoe arrangement of the sarsen Stonehenge trilithons because of their similarity in scale, their enveloping property, the largest stones at the 'back' of the structure and orientated to the south-west, their central location within surrounding stone features, and the fact that they occur in some of the most complex stone circles in Britain.[25] Therefore, by archaeological classification, a condition of identity

22 Ucko et al., *Avebury Reconsidered*, p. 191.

23 Gillings et al., *Landscape of the Megaliths*, pp. 63-90.

24 P.J. Ucko, M. Hunter, A.J. Clark and A. David, *Avebury Reconsidered: from the 1660s to the 1990s* (London: Unwin Hyman, 1991), [hereafter Ucko et al., *Avebury Reconsidered*].

25 R.M.J. Cleal, K.E. Walker and R. Montague, *Stonehenge in its Landscape: twentieth century excavations* (London: English Heritage, 1995); Gillings et al. *Landscape of the Megaliths*.

is claimed for these structures mainly by the property of a closely organised 'quadrangular' arrangement of stones.

Archaeologists have suggested four theories for coves:

1. Seclusion devices nested within the deepest space of surrounding stone circles in which only a select few would have been allowed to enter, and within which rituals would have taken place.[26]
2. Stone facsimiles of prehistoric dwellings, in which the cove represents the dwelling's hearth and the surrounding stone circle representing the walls.[27]
3. The Longstones Cove is a terminal marker for the Beckhampton Avenue.[28]
4. Sacred space to be inhabited by non-corporeal entities, and from which living humans would have been excluded.[29]

As the three Avebury coves belong to the same monument complex but their design and placement vary, interpretation can be reduced to what is common to all three of them. If there is no common element, then we have weakened the archaeological claim of identity. The first theory may be relevant for the cove at the centre of the northern inner circle, since it was surrounded by two stone circles, and these would have created a strong seclusion effect within this mass of stones. However, it cannot explain the placement of the Longstones or West Kennet Avenue Coves, both of which are located within and as part of the two stone avenues.

A similar point weakens the suggestion that coves are facsimiles of Neolithic dwellings, since no circles of megaliths surround these two avenue coves. The dwelling perspective is also severely weakened by the fact that we have very little evidence for any dwellings at all in Avebury, and hardly any in the whole of the Neolithic in the British Isles. When we do have any evidence, it is then of square houses and round hearths — not round houses and square hearths.[30]

The third interpretation that the Beckhampton Cove is a terminal marker for the 'end' of the Beckhampton avenue begs the question of why

26 J. Barrett, *Fragments from Antiquity: an archaeology of social life in Britain, 2900-1200BC* (Oxford: Blackwell, 1994), pp. 17-18; Burl, 'Coves'; Thomas, *Understanding the Neolithic*, p. 214.

27 I. Hodder, *Symbols in Action* (Cambridge: Cambridge University Press, 1982), pp. 224-6; C. Richards, *Dwelling Among the Monuments: the Neolithic village of Barnhouse, Maeshowe passage grave and surrounding monuments at Stennes, Orkney* (Cambridge: McDonald Institute for Archaeological Research, 2005), pp. 218-25.

28 Gillings et al., *Landscape of the Megaliths*, p. 71.

29 Gillings et al., *Landscape of the Megaliths*, p. 168.

30 M. Parker-Pearson, *Stonehenge Riverside Project: New Approaches to Durrington Walls*, http://www.shef.ac.uk/archaeology/research/stonehenge/intro.html [accessed 31 July 2009].

this form of building is needed to mark a terminus, when the same form also marks the centre of the Avebury northern inner circle while the different form of the Sanctuary, a complex stone and post circle, marks the start/end of the West Kennet Avenue. And as there are good reasons to suspect that the Beckhampton Avenue did not terminate at this position, this claim seems to be loading the Longstones Cove with a meaning external to its own design.[31]

That a cove might have been a place for 'spirits' or other non-corporeal beings may well seem to fit the first model, with the cove as seclusion 'chamber' within nested stone circles, but not for a cove to be located along an avenue with an open side, as was the case as reported by Stukeley with the West Kennet Cove and as was probably the case for the Longstones Cove. It is far more probable that a cove added some property to the requirements of rituals within an avenue *and* a circle. However, since we know that the soil surface along and *outside* the West Kennet Avenue was more compressed than the soil surface within it,[32] then this suggests that the structures were not shunned by the living once they had been built.[33]

None of these interpretations can convincingly explain either the form or the placement of coves. This leaves two possibilities — either the category of 'cove' is wrong, or a cove requires other interpretations. The suggestion that coves are seclusion devices, whether or not they are surrounded by stone circles, neglects a paradoxical property — they are open at their corners. The corner gaps are substantial — in the order of 2-3 metres for the Avebury Cove in the inner northern stone circle, and 5-6 metres for the Longstones Cove — and would have provided minimal seclusion properties. If we are claiming 'seclusion' why have open corners, especially if it is just a three-sided 'box'? Therefore, at best, this suggestion can only be a very partial explanation. Nevertheless, the size of these stones is significant — all three coves are made from very large stones, if not the largest in the entire Avebury complex. They are therefore clearly marking these spaces and their architecture to be prominent in some way.

A third property has also been understated — shape. All three stones of the Avebury Cove had straight, vertical sides, flat faces, and lozenge or half-lozenge tops. Aubrey's drawing of the Longstones Cove showed them as having straight vertical sides,[34] and the one remaining 'Adam' stone is similar to the back-stone of the Avebury cove — a massive stone with straight vertical sides with a half-lozenge top. Gaps with straight vertical sides, rather than a sealed enclosure, are one of the design features of these

31 L.D. Sims, 'The Logic of Empirical Proof: a note on the course of the Beckhampton Avenue', *Time and Mind: The Journal of Archaeology, Consciousness and Culture*, (2009), Vol. 2(3), pp. 333-46.

32 Gillings et al., *Landscape of the Megaliths*.

33 Those who walked within the avenue may have been about to become spirits.

34 Gillings et al., *Landscape of the Megaliths*, Fig. 2.63, p. 85.

coves. Rather than considering a cove as a *failed* box, these properties suggest that they are *successful* framing devices, just as the grand trilithon is at Stonehenge.[35]

This suggests a fourth property — the plan layout of the Longstones Cove is not quadrangular but trapezoid. The trapezoid arrangement focuses the side stones on the gaps at the corner edges, and adds to the regularity and fidelity of the corner gaps rather than trying to compensate a loosely organised box. This observation may strengthen the earlier parallel drawn with the Stonehenge trilithons — the two stones L16 and L14 (Fig. 2.1) funnel the eye from the south-east to the edges of L15, just as the trilithons draw the eye from the Heel Stone to the grand trilithon when viewing from the Heel Stone.

Three further properties of coves have been understated in the interests of classifying them as a separate class or category of monument — their placement within other structures is systematic. The Longstones Cove is placed within the Beckhampton Avenue in a position consistent with the average spacing of all the stones in the avenue. Stukeley's drawing of the West Kennet Cove also places that cove as not separate from the avenue, but integrated within it as another, albeit elaborated, part of the avenue.[36] The Avebury coves are not just arithmetical, quantitative elaborations of other structures in the complex, but they are also geometrically related to them. The Avebury Cove is at the centre of the northern inner circle, and the Longstones Cove is integrated into the western row of the Beckhampton avenue while its axis is orthogonal to the axis of that avenue, and similarly Stukeley reported that the West Kennet Cove was also aligned at right angles across the line of the avenue, although now integrated into its eastern row. A circle is defined by its centre and radius, and a line is defined by its length and alignment. All three Avebury coves systematically address by amplification and elaboration the arithmetic and geometric properties of circles and lines. And, lastly, each of these coves are also located in carefully chosen landscape positions — all of them are located at the top of ridges either side of which the land falls away, and while the two avenue coves are on different sides of their respective avenue rows they are both opposite and 'facing' a local, close and high horizon.

From these three Avebury coves, the design features found so far are:

1. A tightly organised group of stones;
2. In a quadrangular or trapezoidal plan arrangement;
3. With stones selected for straight vertical sides;
4. And 'flat' faces;
5. With substantial gaps between the 'corners';
6. An axial orientation;

35 North, *Stonehenge*; Sims, 'Solarization of the Moon'.
36 Ucko et al., *Avebury Reconsidered*, Plate 61, p. 191.

7. In arithmetic dialogue with adjacent megalithic structures;
8. In geometrical dialogue with adjacent megalithic structures;
9. The combination of all the above features — flat faces, vertical sides in a tightly organised space viewed from positions within other adjacent structures — 'pinches' precise gaps at the line of sight intersections of these 'corner' stones;
10. Positioned on local rises in the landscape

Of these ten observable features of the three Avebury coves, only the first two were 'noticed' by any of the archaeological theories. This suggests a post-modern critique that the dominant practice within archaeology is to classify prehistoric structures into types of building so that they may be bracketed with other similar structures elsewhere as precedents or facsimiles. From this study of the Avebury coves, it can be seen that separating off these coves from their context throws away elements of their design which connect them to their *adjacent* and *different* structures. If we include these other properties, then a cove is a *component* structure, not a type structure which stands on its own separate from its context.

Post-modernism usefully, therefore, can demonstrate how a disciplinary practice constructs classifications that ignore certain features of a thing. On the other hand, it does not sensitise us to the different theories and methods that can engage with a fuller list of observations. We have unexplained data for the Avebury coves that is considered 'enigmatic' by archaeology, and which appear to be design features of the cove structures we have considered so far. If we can find another theory or method that notices all ten and more of these features, then that theory is stronger than those that cannot. If this exercise is successful then it also weakens the post-modern critique of realist scholarship.

An enacted cosmology is a multi-media ritual event. It would include not just depositing selected and processed items of material culture beneath and around built structures, later to be excavated by field archaeology, but the design of these structures would also have aspired to bringing all spheres of their worlds into some sort of coherence — a cosmology. This would have included the above and below this-world transit of cosmic bodies, since all cosmologies and religions share the stricture — as it is in heaven, so it is on earth. Gillings et al. claim that no convincing demonstration has been made for astronomical alignments at coves.[37]

Before a judgement of 'non-convincing' can be accepted, scholarship expects that this claim is demonstrated by a critique of those who *have* suggested 'astronomical' alignments at coves. Since many archaeologists until recently have not been convinced by the entire discipline of archaeoastronomy, then it may be appropriate to adopt a post-modern suggestion that the professional reticence of archaeology encourages

37 Gillings et al., *Landscape of the Megaliths*, p. 169.

scepticism that is above and beyond scholarship, and that this position is driven largely by the perceived political requirements of a boundary dispute. Three scholars have made such claims – Ruggles,[38] Burl[39] and North.[40] Ruggles points out that coves exhibit a scatter of orientations on lunar standstill and solstice horizon events, and both Burl and North have suggested that the Avebury Cove has lunar-solar alignments built into its design.

Let us first consider North's claim. According to North the Avebury Cove is the focus for two lunar and one solar ray observable from the outer ring of stones looking into centre of the northern inner circle along the tangents of intervening circular structures (Fig. 1.2). As both of the centres of the two inner circles stand on an undulating ridge that runs roughly north-south through the main circle, all of these rays when viewed from the outer circle trace their path *uphill*, so framing a clear view of the sky behind – not the landscape. The stones of the outer circle are numbered 1-99 clockwise from the south-eastern entrance. It can be seen from Fig. 1.2 that from stone 19 in the outer circle, a ray traces its way through the left hand gap in the back of the cove, touching the edge of the inner circle of the double post circle and then above the outer northeast bank to the northern major moonrises.[41]

Moving around the outer circle to the position of stone 65, a ray touches the left hand side of the outer circle of the double post circle, threads its way through the right hand gap at the back of the Cove and then above the outer southwest bank to the winter solstice sunsets. Then around to the position of stones 89-90 of the outer circle another ray at right angles to the winter solstice sunset ray passes the right hand side of the southern inner circle and then along the inner face of the back stone of the Cove, beyond the outer bank in the northwest, towards the summer solstice sun setting into the summit of Windmill Hill. Notice that all three inner faces of the Cove have alignments on sun and moon and are arranged either in reverse or orthogonally to each other. These are all claims that can be tested by fieldwork, virtual reality modelling, and Monte Carlo random modelling.[42]

Also, just as at Stonehenge when walking towards the grand trilithon from the Heel Stone, when walking uphill towards the Cove framing the winter solstice sunset, the upward motion of the observer's eye counter-

38 Statement by Ruggles in Villicana, *Celebrating the Summer Solstice.*

39 A. Burl, *Prehistoric Avebury* (London: Yale, 2002), p. 147.

40 North, *Stonehenge*, pp. 271-6.

41 Lunar standstills occur twice every nineteen years and are each spread over the course of about one year. The human eye cannot differentiate the horizon positions of solstice sunrise and sunset alignments for three days either side of the solstice. Therefore, there are about thirteen moonsets every sidereal month that define a lunar standstill and seven days of solstice by unaided eye horizon 'astronomy' (Sims, 'Solarization of the Moon').

42 Macdonald 2009; Ruggles, *Astronomy in Prehistoric Britain and Ireland.*

balances the apparent setting motion of the sun to create the illusion of holding 'time' still.[43] If this is not an over-interpretation, we would expect to find similar and consistent properties with the Longstones and West Kennet Avenue coves. Turning to the Longstones Cove in Fig. 4 it can be seen that as a component of the Beckhampton Avenue, using adjacent avenue stones as back-sights and fore-sights, reversible and orthogonal alignments exist through the corner gaps of the Longstones Cove which continue on to adjacent avenue stones. By combining the published site excavation plan alignments with independent field survey of the Adam stone,[44] we can minimise some of the errors inherent in this exercise.

The results are surprisingly in keeping with what our model predicts. The ray to the northeast which threads through the Cove following the line of the western row of the avenue is aligned on the northern major standstill moonrises, the southern major standstill moonsets in reverse, and at right angles to that line a ray continues at an altitude of nearly 3° to the summit of Folly Hill, where was once the barrow A2,[45] and aligns on the winter solstice sunrise.

Therefore, at both the Avebury Circle Cove and the Longstones Cove we have found double horizon alignments on the solstice sun and the standstill moon. These combinations invariably and predictably generate a dark moon coinciding with both the winter and summer solstice sun.[46]

We could have generated the same finding as *predictions* from lunar-solar conflation theory.[47] It is a paradox of the Avebury monument complex that the two avenues' routes allow only partial views of Silbury Hill from just five positions, otherwise obscuring all view of Silbury Hill for nearly 80% of their length. Viewing the scoured chalk of the summit platform of Silbury Hill from Fox Covert, the Beckhampton Avenue crossing of the River Winterbourne, the centre of the inner southern circle of Avebury circle, and the Sanctuary, simulates seeing the whitish summit platform as crescent moon's first or last glint before and after dark moon at winter solstice.[48] From two of these positions, at the River Winterbourne and the Sanctuary, the level summit platform of Silbury Hill is exactly in line with the background horizon. For a viewer who sees Silbury Hill as the moon, by the artifice of building it in the local landscape in line with the background horizon, it would signify to them that the moon had set. To see the moon when it has set, then, by your own agency this is only possible if you

43 North, *Stonehenge*; Sims, 'Solarization of the Moon'.

44 See L14 in Gillings et al., *Landscape of the Megaliths*, Fig. 2.83, p. 125; with thanks to Steve Marshall — personal communication.

45 A.B. Powell, M.J. Allen and I. Barnes, *Archaeology in the Avebury Area, Wiltshire: Recent Discoveries Along the Line of the Kennet Valley Foul Sewer Pipeline, 1993* (Wessex Archaeology: Report No. 8, 1996), p. 14.

46 Sims, 'Lunar Standstill'.

47 Sims, 'Solarization of the Moon'.

48 Sims, 'The Underworld'.

yourself are in the underworld. The monument complex is therefore designed to simulate a route through the underworld.

The Beckhampton Avenue 'starts' at Fox Covert moving towards the Longstones Cove with a view of Silbury Hill proud of the eastern horizon (Fig. 5). Only the waning crescent moon can be seen rising on the eastern horizon, since it is just to the right of the soon to be risen sun still below the horizon.

Once the sun has risen, its light outshines the light of the waning crescent and it can no longer be seen for any of its daytime transit across the sky. Contrarily, the waxing crescent moon is to the left of the sun, and can only be seen setting low on the western horizon once the sun has already set. Waxing crescent moon cannot be seen rising, since the already risen sun again outshines its light. Therefore, waning crescent moon is associated with the rising sun and waxing crescent moon is associated with the setting sun. Processing along Beckhampton Avenue from Fox Covert, we are therefore being bracketed with the rising waning crescent moon view of Silbury Hill proud of the eastern horizon.

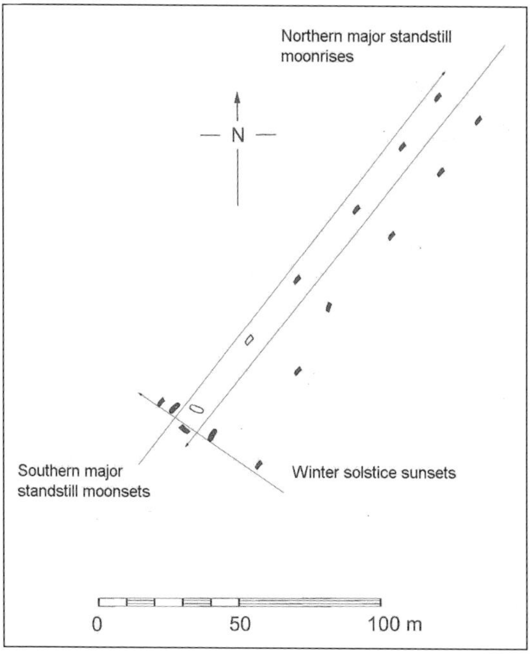

Figure 4: Alignments at the Longstones Cove.[49]

49 Gillings et al., *Landscape of the Megaliths*, p. 63.

We would predict with this model that along this part of the monument complex any solar alignment would be on winter sunrise, consistent with a south eastern horizon viewing of rising waning crescent moon. In keeping with this prediction, we find that at the base of the northeast side of the Adam stone of the Longstones Cove, a human 'burial' of an adult male had his head pointing to the south-east, in line with the alignment of the Cove towards the Folly Hill barrow complex and winter solstice sunrise.

Therefore, both inductively through the data, and deductively from this theory, we converge at the same interpretation of a lunar-solar conflation of alignments at the Longstones cove. When two different procedures and bodies of evidence converge on the same finding, this exponentially raises our confidence in our interpretation of the observations. They become 'facts' for our theory.

We have just Stukeley's record of the West Kennet Cove, but no recovered site plan. However, since we have found that the other two coves at Avebury are integrated into their building and landscape context, we can predict from the avenue remains and surrounding landscape some of the expected properties of this cove. We risk our model to observational test. From those parts of the West Kennet Avenue which remain in the northern section, viewing across it to the southwest over the crest of Waden Hill the West Kennet Avenue is designed to see the winter solstice sunset across paired stones, and southern minor standstill moonsets across one set of diagonals and cardinal alignments across the other diagonal.[50] All five burials found along this section of the Avenue are, like that at the Longstones Cove, also on the northeast sides of the stones, but for the one of which we have information, his head points to the southwest, consistent with the stone's alignment on winter solstice sunset.

We would expect the West Kennet Avenue Cove, once rediscovered and surveyed, to frame that event with more accuracy than possible with just paired stones and to repeat with greater fidelity the alignments along the avenue to the southeast to frame the rising southern minor standstill moonrises. However, while these are predictions testable by a future site excavation, there is one property we can test now without excavation of the West Kennet Avenue Cove — its horizon views from its known position. Stukeley said that the West Kennet Cove stood at the 'apex' to the avenue, and showed that it would have been at position 50a along the modern numbering system for the avenue.[51]

This part of the landscape just next to the modern road is a flat saddle between two gentle gullies in the western side of the dry valley. At right angles to the line of the avenue and on the high horizon to the east lies a very large barrow marking the sky-line. But this position, in the middle of

50 L.D. Sims, 'Gender, Power and Asymmetry in the Neolithic: the West Kennet Avenue, Wiltshire, as a test case' (forthcoming).
51 Fig. 3 and Ucko et al., *Avebury Reconsidered*, p. 190.

the West Kennet Avenue, would have been the first position coming from the Sanctuary where it would have been possible to see the Avebury circle outer bank with the top of the tallest stone within the southern inner circle, the 'Obelisk', protruding above it (Fig. 3). Looking back to the Sanctuary – a complex multiple circle of stones and lintelled posts at the start/terminus of the West Kennet Avenue (Fig. 5) – the winter solstice sun would have risen out of its top.

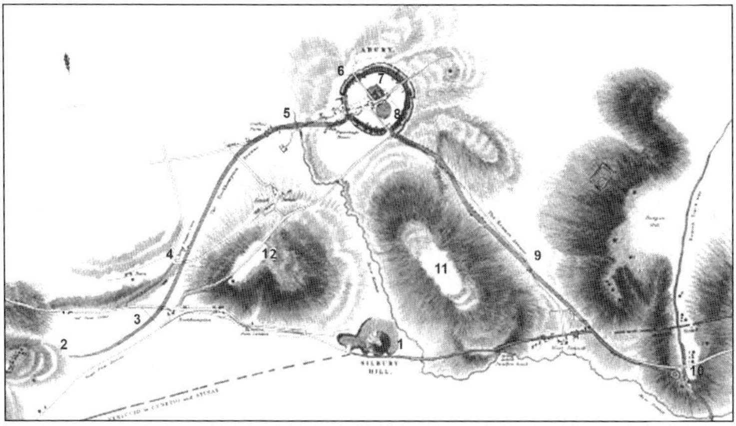

Figure 5: The Avebury complex.[52]
Key: 1. Hill 2. Fox Covert 3. Beckhampton Avenue 4. Longstones Cove 5. River Winterbourne 6. Avebury circle and henge 7. Northern inner circle 8. Southern inner circle 9. West Kennet Avenue 10. Sanctuary 11. Waden Hill 12. Folly Hill.

Therefore, just as the Cove at the centre of the northern inner circle of the henge manipulates an alignment on the summer solstice sunset to also set in the summit of the causewayed enclosure on the summit of Windmill Hill, so the placement of the West Kennet Cove is in dialogue with views to the northwest towards Avebury circle and to the southeast towards the Sanctuary coinciding with winter solstice sunrise. It is located at the tipping point between a dark moon ritual at winter solstice sunset at the circle and winter solstice sunrise at the Sanctuary. Coves therefore act as focussing devices for horizon 'astronomy', bringing sky and landscape together as a coupled system. Therefore, we may add to the list of features for coves:

1. Their position in a monument complex is selected to facilitate an arrangement of stones for high fidelity reversible and cruciform lunar-solar alignments;

52 Image adapted from Pete Glastonbury, *Avebury Panoramic Tour*, CD ROM (2001).

2. These alignments follow a winter solstice lunar scheduling before, during and after a dark moon ritual;
3. They are located in a landscape position which manipulates the horizon viewing of other structures of different materialities (the chalk and wood of Windmill Hill, mainly chalk, stone and a little wood of Avebury circle, and the equal quantities of stone and wood of the Sanctuary).

Ruggles sees coves' multiple alignments as lacking a 'discernable commonality'.[53] Instead, these findings suggest that a dark moon ritual at winter solstice requires a 'grammar' of different alignments for the enactment of an initiatory myth of journeying through the underworld during the longest, darkest night between winter solstice sunset and sunrise. 'Astronomy' and landscape are integrated as a coupled system to facilitate this 'grammar'. Ruggles sees cove alignments as either not stabilising around a single cosmically relevant value or as being aligned on local landscape features distinct from the horizon rise/set positions for the sun or the moon. For Ruggles, this severely limits the archaeoastronomical method. We have found that the Avebury coves do both. This could only be a problem if the assumption were being made that being a 'single' type of structure they 'should' all share the same single alignment. This assumption is consistent with the present archaeological practice of classifying different structures as 'types', which share a common identity, and in the present archaeoastronomical use of the statistical method, which seeks a class of monuments of the same design and expects to 'prove' archaeoastronomy by finding the same alignments for all of them beyond what would be expected by random variation alone. Both disciplinary preoccupations, driven by contemporary political processes internal to the academy, weaken observational and interpretive skills.

However, we have found that coves are probably not type structures distinct from their monument context, but that they are *component* structures best understood by their dialogue with the avenues and circles within which they are located. Therefore, mobilising methods for interpreting unique or idiosyncratic structures are more relevant to studying coves than the statistical method. The statistical method may perhaps have been appropriate for a stage of archaeoastronomy designed to overcome archaeological scepticism to the claims of an earlier stage of the horizon astronomy, but it is not subtle enough to deal with coves. It cannot see most of the defining properties of coves, and reflects a political agenda of establishing disciplinary boundaries in the academy.

The findings we have made at Avebury appear to be typical of the few other coves known.[54] It is also the case that, if we consider the trilithon

53 Statement by Ruggles in Villicana, *Celebrating the Summer Solstice*.
54 It has been reported that the cove at Arbor Low faces the northern major lunar standstill (Alex Whitaker, *Ancient Wisdom* website, http://www.ancient-

horseshoe at sarsen Stonehenge as a type of cove, then the lunar-solar double window there amplifies these findings from Avebury and reduces the number of theories necessary to interpret contemporary monuments of varying designs.

Notice that the full suite of thirteen properties we have found for the Avebury coves will not support a theory such as farming revolution theory which assumes the primitivism of the builders. These structures display a sophisticated syntax of alignments of a lunar-solar and locally marked landscape cosmology. This is consistent with a theory of Palaeolithic/ Mesolithic hunter-gatherer respect for the moon which is being confiscated to the novel purposes of Neolithic/EBA cattle herders. This model also fits with the evidence for lunar-solar conflation already found useful to interpret Stonehenge and Silbury Hill — avenue routes and their component coves follow a lunar-solar logic for staging a dark moon winter solstice ritual at Stonehenge and in the Avebury circle which simulates a journey into and returning from the underworld.[55] Critics of this interpretation, whether post-modern or not, must show either that the data is wrong, or that they have a better explanation of the same data. Neither post-modern abstractions nor limiting coves to just a quadrangular arrangement of closely grouped stones will be good enough to encompass all thirteen properties we have found for the Avebury coves.

Cultural astronomy — a possible future

I have utilised the post-modern critique of scholarship to interpret politico-disciplinary motives current within archaeology and archaeoastronomy which impede the interpretation of coves. But I have also argued that there are severe limits to this perspective which fail to account for the patterns within observations. I have adopted a realist theory of knowledge, in which observations can be invested with interpretive meaning once we mobilise relevant theories. I have used an American definition of anthropology as an integrative discipline that links the life sciences and social/cultural disciplines so that we can ask — what is it cosmology? Every discipline and method from biological anthropology, archaeology, linguistics, social anthropology to myth and folk-lore, can then be called upon to reconstruct and triangulate from ancient fragments and modern varieties of behaviour and culture the lost and obscure evidence of all culture's cosmologies. In this paper I have used archaeology, antiquarian testimony, archaeo-astronomy and anthropology to mobilise an interpretation of different

wisdom.co.uk/englandarborlow.htm [accessed 21 July 2009]), the cove at Stanton Drew had a very rough alignment to the major southern midsummer moonrise (A. Burl, *Great Stone Circles* (London: Yale, 1999), p. 54), Mount Pleasant had lunar-solar cruciform alignments (North, *Stonehenge*, p. 382), and within Stenness Stones circle the cove's two portals align on the round boss of Maes Howe.

55 Sims, 'Solarization of the Moon'; Sims, 'The Underworld'.

types of observations. This multi-disciplinary technique does not just 'triangulate' meaning around a single interpretation, but assuming that each set of observations are correct, it only allows a very limited combination of all these elements which recreate the totality of the cosmology of lost cultures. This is the principle of 'emergence'. In this case, cultural astronomy is central to enhancing the ability of the other methods in rebuilding the lost reality of coves.

Horizon astronomy, which sees alignments as a coupled system of landscape and sky, can provide the missing ingredient which raises the power of the other methodologies to move towards a deeper interpretation of the cosmologies of the past.

Cultural astronomy, therefore, has a key role to play in interpreting how every culture tries to make sense of 'life' through its 'cosmology'. This paper has argued that this role can only be achieved once scholarship is able to transcend both the post-modern critique and the over-narrow definition of field method current within archaeoastronomy. By adopting a multi-disciplinary approach, or an American definition of anthropology, different methodologies can be bought to bear on any culture's cosmology which, when combined, achieve 'emergent' properties which exponentially reduce the possible number of testable interpretations. This methodology has been demonstrated through a new interpretation of 'coves' which is consistent with a recent 'transformational template' of lunar-solar conflation.

Within archaeology, cultural origins are largely assumed to derive from a Neolithic farming revolution.[56] Instead, I have adopted an origins scenario which locates cultural origins amongst low-latitude Palaeolithic hunter-gatherers in Africa. This scenario predicts that the first culture-bearing modern humans synchronised their rituals, and therefore their hunting and politico-sexual lives, according to a lunar time schedule.[57] Lunar-solar conflation theory is derived from a transformation of this original lunar template, and which later takes on properties which alienate and adjust lunar properties to an estranging and authoritarian logic of emerging elite cattle-owning males organising within an emerging solar cult. This transformational template fits the circumstances of the builders of the Avebury monument complex, and points one way forward for cultural astronomy's search for a wider vocabulary to interrogate the concept of 'cosmology'.

Acknowledgements

I would like to thank Steve Marshall for comments on an earlier version of this paper and for generously sharing with me his Longstones Cove field data, as well as John MacDonald for sharing his observations of the shape of the Avebury Cove stones.

56 Thomas, *Understanding the Neolithic.*
57 Knight et al., 'Human Symbolic Revolution'.

Bibliography

Barrett, J., *Fragments from Antiquity: An Archaeology of Social Life in Britain, 2900-1200BC* (Oxford: Blackwell,1994).

Burl, A., *Prehistoric Avebury* (London: Yale, 2002).

Burl, A., *Great Stone Circles* (London: Yale, 1999),

Burl, A., 'Coves: structural enigmas of the Neolithic', *Wiltshire Archaeological and Natural History Magazine,* (1988), Vol. 82, pp. 1-18.

Cleal, R.M.J., K.E. Walker and R. Montague, *Stonehenge in its Landscape: Twentieth Century Excavations* (London: English Heritage, 1995).

D' Errico, F., 'The Invisible Frontier: A Multiple Species Model for the Origin of Behavioral Modernity', *Evolutionary Anthropology,* (2003), Vol. 12, pp. 188-202.

Fountain, J.W. and R.M. Sinclair, *Current Studies in Archaeoastronomy: Conversations Across Time and Space* (Durham: Carolina Academic Press, 2005).

Gillings, M., J. Pollard, D. Wheatley and R. Peterson, *Landscape of the Megaliths: Excavation and Fieldwork on the Avebury Monuments, 1997-2003* (Oxford: Oxbow, 2008).

Glastonbury, P., *Avebury Panoramic Tour,* CD ROM (2001).

Henshilwood, C. and C. Marean, 'The Origin of Modern Human Behaviour: Critique of the models and their test implications', *Current Anthropology,* (2003), Vol. 44(5), pp. 627-651.

Hodder, I., *Symbols in Action* (Cambridge: Cambridge University Press, 1982).

Hoskin, M., *Tombs, Temples and Their Orientations: A New Perspective on Mediterranean Prehistory* (Bognor Regis: Ocarina, 2001).

Knight, C., *Menstruation and the Origins of Culture* (London: University College London [unpublished PhD thesis], 1987).

Knight, C., *Blood Relations* (New Haven: Yale University Press, 1995).

Knight, C., C. Power & I. Watts, 'The Human Symbolic Revolution: a Darwinian Account', *Cambridge Archaeology,* (1995), Vol. 5, pp. 75-114.

Levi-Strauss, C., *The Naked Man: Mythologiques, Volume 4,* trans. John and Doreen Weightman (Chicago: Chicago University Press, 1990).

Levi-Strauss, C., *The Elementary Structures of Kinship* (Boston: Beacon Press, 1969).

North, J., *Stonehenge: Neolithic Man and the Cosmos* (London: Harper Collins, 1996).

Macdonald, J., *Stonehenge 3D* website, http://www.stonehenge3d.co.uk/ [accessed 10 August 2009].

Parker-Pearson, M., *Stonehenge Riverside Project: New Approaches to Durrington Walls,* http://www.shef.ac.uk/archaeology/research/stonehenge/intro.html [accessed 31 July 2009].

Powell, A.B., M.J. Allen and I. Barnes, *Archaeology in the Avebury Area, Wiltshire: Recent Discoveries Along the Line of the Kennet Valley Foul Sewer Pipeline, 1993* (Wessex Archaeology: Report No. 8, 1996).

Richards, C., *Dwelling Among the Monuments: the Neolithic village of Barnhouse, Maeshowe passage grave and surrounding monuments at Stennes, Orkney* (Cambridge: McDonald Institute for Archaeological Research, 2005).

Ruggles, C., *Astronomy in Prehistoric Britain and Ireland* (London: Yale, 1999).

Ruggles, C., Statement made by Clive Ruggles in the film *Celebrating the Summer Solstice: The Pagan Experience,* by Darlene Villicana, (2009), shown at the Sophia Centre 'Cosmologies' Conference, 6 June 2009, in Bath, UK.

Sims, L.D., 'The Solarization of the Moon: manipulated knowledge at sarsen Stonehenge', *Cambridge Archaeological Journal,* (2006), Vol. 16(2), pp. 191-207.

Sims, L.D., 'What is a Lunar Standstill? Problems of Accuracy and Validity in the Thom Paradigm', *Mediterranean Archaeology & Archaeometry,* (2007), Vol. 6(3), pp. 157-163.

Sims, L.D., 'Entering, and Returning from, the Underworld: Reconstituting Silbury Hill by Combining a Quantified Landscape Phenomenology with archaeoastronomy', *Journal of the Royal Anthropological Institute,* (2009), Vol. 15(2), pp. 386-408.

Sims, L.D., 'The Logic of Empirical Proof: A Note on the Course of the Beckhampton Avenue', *Time and Mind: The Journal of Archaeology, Consciousness and Culture*, (2009), Vol. 2(3), pp. 333-46.

Sims, L.D., 'Gender, Power and Asymmetry in the Neolithic: the West Kennet Avenue, Wiltshire, as a test case' (forthcoming).

Thomas, J., *Understanding the Neolithic* (London: Routledge, 1999).

Ucko, P.J., M. Hunter, A.J. Clark and A. David, *Avebury Reconsidered: from the 1660s to the 1990s* (London: Unwin Hyman, 1991).

Whitaker, A., *Ancient Wisdom* website, (2009), http://www.ancientwisdom.co.uk/englandarborlow.htm [accessed 21 July 2009].

Calendars and Divination in the Dead Sea Scrolls: the Case of 4Q318: 4QZodiac Calendar and Brontologion

Helen R. Jacobus

This Dead Sea Scroll, written in Aramaic, has two units: a zodiac calendar and a thunder omen text (brontologion). It will be shown that the zodiac calendar is mathematically sophisticated. Its accompanying thunder omen text has standard Hellenistic parallels. The genre belongs to the late Graeco-Roman period and there is an identifiable Mesopotamian background. This scroll is important for the study of the development of the Jewish calendar, the history of astronomy, astrology, and ancient science.

The corpus of mainly Hebrew texts found in the caves at and around Qumran, by the Dead Sea, known as the Dead Sea Scrolls, consists of the remains of a large body of writings by different Jewish groups who flourished in Israel around 2,000 years ago.[1]

There are a substantial number of biblical texts from which scholars are currently learning about the textual development of different books in the Hebrew Bible.[2] The archives of those who wrote or preserved the scrolls also contain a variety of non-biblical texts, these include prayers, biblical commentaries, rules of a community who may have lived at Qumran or belonged to its sect, different kinds of calendars, and two texts which explicitly use the zodiac.[3]

1 Philip R. Davies, 'Sect Formation in Early Judaism', in David J. Chalcraft, ed., *Sectarianism in Early Judaism* (London: Equinox, 2007), pp. 133-55; Geza Vermes, *The Complete Dead Sea Scrolls in English* (London: Penguin, 1997), pp. 26-90.

2 For a translation of the biblical scrolls for the general reader, see: Martin Abegg, Peter Flint and Eugene Ulrich, eds., *The Dead Sea Scrolls Bible: The Oldest Known Bible Translated for the First Time into English* (New York: HarperCollins, 1999).

3 For the most current translation of the non-biblical scrolls for general readers, see: Michael Wise, Martin Abegg and Edward Cook, eds., *The Dead Sea Scrolls: A New Translation* (New York: HarperCollins, 2005), [hereafter Wise et al., *Dead Sea Scrolls*]; Matthias Albani, 'Horoscopes', in L.H. Schiffman and J.C. VanderKam, eds., *Encyclopedia of the Dead Scrolls* (Oxford: Oxford University Press, 2000), [hereafter Albani, 'Horoscopes'], pp. 370-73; Mathias Albani, 'Horoscopes in the Qumran

One of these is a zodiacal physiognomic text in Hebrew, written in code;[4] and the other is lunar zodiac calendar with a divination text, catalogued as 4Q318 (4Q stands for Qumran Cave 4) with the title 4QZodiology and Brontology ar ('ar' denotes that the text is written in Aramaic);[5] this author prefers the title 4QZodiac Calendar and Brontologion, an adaptation of that assigned to the text by Geza Vermes.[6]

Description of 4Q318

The portable, fragmentary, parchment manuscript is dated to around the end of the first century BCE to the beginning of the first century CE.[7] It is a small scroll consisting of a writing block 11.5cm wide and 8.5cm deep. The surviving columns of Aramaic consist of two units: a zodiac calendar with a brontologion, that is, a thunder omen text (from the Greek *brontos*, 'thunder', and *logion*, 'speech'). Probably about one-sixth of the original is extant: just over two columns remain from a possible eleven or twelve-column manuscript.

These extant, lacunose columns of writing contain the last two and a half months of the year, and the beginning of the brontologion. The latter consists of prognostications for two signs of the zodiac, using a protasis-apodasis formulaic structure (if *x*, then *y*), familiar from both Mesopotamian omina and Hellenistic divination texts.[8]

The calendar lists each day of the month and the moon's diurnal position in the zodiac. The user of this manuscript would apply the calendar to work out the zodiac sign of the moon on the date when the thunder-clap takes place. It would then be possible to give an interpretation for the meaning of the thunder according to the written prediction for that sign. When restored, it can be seen that the calendar is schematic and begins with the moon in Taurus on days one and two of the first month, Nisan (the Aramaic translation of the Babylonian month-name, Nisannu),[9] see Table 1.

Scrolls', in J.C. Vanderkam and P.W. Flint, eds., *The Dead Sea Scrolls After Fifty Years* (Leiden: Brill, 1999), pp. 279-330.

4 M. Wise, 'A Horoscope Written in Code (4Q186)', in Wise, Abegg and Cook, *Scrolls*, pp. 275-78; Mladen Popović, 'Reading the Human Body and Writing in Code: Physio-nomic Divination and Astrology in the Dead Sea Scrolls', in Anthony Hilhorst, Émile Puech and Eibert Tigchelaar, eds., *Flores Florentino: Dead Sea Scrolls and Other Early Jewish Studies in Honour of Florentino García Martínez* (Leiden: Brill, 2007), pp. 271-84.

5 J.C. Greenfield and M. Sokoloff, with A. Yardeni and D. Pingree, '318. 4QZodiology and Brontology ar', in Pfann et al., eds., *Discoveries in the Judaean Desert. Volume 36: Qumran Cave 4: XXVI* (Oxford: Clarendon Press, 2000), [hereafter Greenfield et al., '318. 4QZodiology and Brontology ar'], pp. 259-72, plates 15-16.

6 Vermes, *The Complete Dead Sea Scrolls in English*, p. 361.

7 A. Yardeni, in Greenfield et al., '318. 4QZodiology and Brontology ar', p. 260.

8 F. Rochberg, *The Heavenly Writing: Divination, Horoscopy, and Astronomy in Mesopotamian Culture* (Cambridge: Cambridge University Press, 2004), pp. 4-13, 39-97, 55.

9 E.J. Bickerman, 'Calendars and Chronology', in W.D. Davies and L. Finkelstein, eds.,

The Aramaic month-names are still in use in the Jewish, luni-solar calendar today.

The lunar month of Nisan overlaps with the sun in Aries, depending on where the conjunction takes place. In this text, the moon is schematically at the beginning of Taurus on the first day of the month; therefore, the conjunction could possibly occur at about 18° Aries.[10] The moon's phase at the beginning of the month would be, therefore, the first visible lunar crescent, as it is in the Mesopotamian luni-solar calendar.[11] (Unfortunately, Wise dismissed this reasoning because he thought that the text described a 364-day year, in common with the calendars of the priestly courses at Qumran).[12]

The Brontologion

The brontologion in 4Q318 reflects the accompanying zodiac calendar by also beginning in Taurus. Below is a translation of the remains of the thunder omen text, written on the last three and a half lines of the manuscript (4Q318 col 8, lines 6-9).

> 6. *vacat* [If in Taurus] it thunders (there will be) *mtsbt* against [
> 7. [and] affliction for the province, and a sword [in the cou]rt of the king and in the province, [
> 8. will be. And to the Arabs [ ], hunger, and they will plunder each oth[er *vac*]*at*
> 9. If in Gemini it thunders, (there will be) fear and sickness from the foreigners and *m*[[13]

Cultural background to the Brontologion

As the first scholars who studied this text — Milik, Wise, Greenfield and Sokoloff with Pingree, and Albani — have pointed out, the Qumran thunder-omen text is virtually identical to late Hellenistic brontologia, versions of which have survived in late Byzantine, medieval and other late secondary sources.[14] Albani, Pingree and Wise observed that there is a structural

---

The Cambridge History of Judaism, Volume 1 (Cambridge: Cambridge University Press, 1984), pp. 60-69.

10 See also M.O. Wise, 'Thunder in Gemini: An Aramaic Brontologion (4Q318) from Qumran', in M.O. Wise, ed., *Thunder in Gemini* (Sheffield: Sheffield Academic Press, 1994), [hereafter Wise, 'Aramaic Brontologion'), pp. 40-42.

11 A.J. Sachs and H. Hunger, *Astronomical Diaries and Related Texts from Babylonia, Volume 1: Diaries from 652BC to 262BC* (Vienna: Verlag der Österreichischen Akademie der Wissenschaften, 1988), [hereafter Sachs and Hunger, *Astronomical Diaries*], p. 20.

12 Wise, 'Aramaic Brontologion', p. 42.

13 Greenfield et al., '318. 4QZodiology and Brontology ar', p. 264.

14 J.T. Milik, *Ten Years of Discovery in the Wilderness of Judea*, J. Strugnell, trans. (London: SCM, 1959). [hereafter Milik, *Ten Years of Discovery*], p. 42; D. Pingree, in Greenfield et al., '318. 4QZodiology and Brontology ar', pp. 270-72; Wise, 'Aramaic

parallel between 4Q318 and Suppl. gr. 1192 folios 42v to 47, in the Bibliothèque Nationale in Paris.[15] This sixteenth-century rendering of an apparent Hellenistic-style 'selenodromion' (*selene*: 'moon', *dromologia*: 'timetable'), a lunar zodiac calendar with an accompanying zodiacal brontologion, is one of several similar texts published in the monumental opus of Greek astrological manuscripts, *Catalogus Codicum Astrologorum Graecorum.*

Pingree, Wise and Albani further noted the textual similarity between the 4Q318 brontologion and earlier Mesopotamian *omina* (in particular, Pingree singled out the [still] unpublished *Enūma Anu Enlil*, tablet 44). Wise compared the 4Q318 brontologion with numerous Hellenistic and Akkadian omen texts throughout his study.

Milik drew attention to the parallel between the Aramaic brontologion from Qumran and the tenth century compendium of practical, rural folklore, *Geoponica* Book 1, chapter 10.[16] The few Hellenistic astronomical divination texts in the *Geoponica* include predictions about king and country commodity prices and the weather from the position of the moon and Jupiter in the zodiac.[17] It also contains presages for the year ahead, based on the moon's place in the zodiac when the Dog Star, Sirius, rises heliacally.[18]

Geoponica further describes prognostications for the forthcoming year, determined by the moon's position in the zodiac when the first thunder occurs after the heliacal rising of Sirius (*Geoponica*, 1:10 is attributed to 'Zoroastres'[19]). The parts of this section when the moon is in Taurus and Gemini,[20] in particular, have striking similarities with the meagre remains of the 4Q318 brontologion, for which only Taurus and Gemini are attested. Both *Geoponica* 1:10 vss 3-4 and 4Q318 col. 8 lines 6-8 refer to the royal court and the Arabs within those two zodiac signs. Neither of these elements occurs in any of the other zodiac signs in the *Geoponica* text.

Brontologion', pp. 13-50; M. Albani, 'Der Zodiakos in 4Q318 und die Henoch-Astronomie', in *Mitteilungen und Beiträge der Forschungsstelle Judentum der Theologischen Fakultät Leipzig,* (1993), Vol. 7, [hereafter Albani, 'Der Zodiakos in 4Q318'], pp. 4-20.

15 P. Boudreaux, ed., *Catalogus Codicum Astrologorum Graecorum*, 8.3 (Brussels: Lamertin, 1912), [hereafter Boudreaux, *Catalogus Codicum,* 8.3], f.42v to 47, pp. 193-97. See M. Albani, 'Der Zodiakos in 4Q318', p. 17, n. 44; D. Pingree, in Greenfield et al., '318. 4QZodiology and Brontology ar', p. 271; Wise, 'Aramaic Brontologion', pp. 27, 29, 31, 33-35, and notes, 36, 48, 63, 69, 77, 78.

16 H. Beckh, ed., *Geoponica sive Cassiani Bassi Scholastici de re Rustica Eclogae* (Leipzig: Biblioteca Teubneriana, 1895), [hereafter Beckh, *Geoponica sive Cassiani*]; Thomas Owen, trans., *Geoponica: Agricultural Pursuits* (London: Spilsbury, 1805-6). [hereafter Owen, *Geoponica*]; Milik, *Ten Years of Discovery*, p. 42.

17 For forecasts concerning Jupiter through the signs of the zodiac, see Owen, *Geoponica*, 1:12, pp. 22-30.

18 Owen, *Geoponica* 1:8, pp. 16-17; Beckh, *Geoponica sive Cassiani,* pp. 15-17.

19 Owen, *Geoponica*, x, xi; 1:10, pp. 19-21.

20 Beckh, *Geoponica sive Cassiani*, p. 19; T. Owen, *Geoponica,* 1:10, p. 19.

The relevant extract (with similarities to the Qumran brontologion, in bold) is as follows:

> *If it thunders when the moon is in Taurus,* it is a sign that the wheat and barley will be injured, and that there will be affliction from locusts but mirth *in the royal palace* ... and to them in the east, vexation and famine. *If it thunders when it is in Gemini,* it portends trouble and disease, and injury to the corn, and *perdition to the Arabs* ᾿Αράβων ...[21] [Emphases added].

Interestingly, Boudreaux also drew a comparison between *Geoponica* 1:10 and Paris Suppl. gr. 1191[22] more than forty years before Milik revealed the existence of the parallel brontologion in the Dead Sea Scrolls.

A previously unpublished brontologion: the prognostication for thunder when the moon is in Aquarius, has recently been discovered by Daryn Lehoux.[23] Excised from the modern scholarly publication,[24] it is combined with the month of February in a version of a Graeco-Roman cultic calendar, for one year, in the manuscript known as the Oxford Parapegma. This thirteenth-century copy of an earlier text that contains cultic information about feast-days, includes the date for the Egyptian New Year — on 20 August in the Egyptian 'wandering' year — and a reference to the birthday of Augustus on 23 September.

It is evident from this data that the original must have been composed very shortly before the death of Augustus in 14CE because the wandering Egyptian New Year date coincides with the year ahead, 15CE.[25] This means that the Oxford Parapegma would be contemporaneous with the paleographical dating of 4Q318 by Yardeni, although it is not known when the brontologion was added.

The zodiac calendar

4Q318 is the largest fragment at Qumran with extant Aramaic month names (Shevat, month 11, and Adar, month 12, survive). It is here suggested that 4Q318 may be a pre-rabbinical, proto Jewish calendar. The results of the empirical tests, below, suggest that the calendar works on the same Metonic cycle as the rabbinical Jewish calendar, in which nineteen solar years are coordinated with 235 synodic lunar months, a system derived from Babylonia.

---

21 Owen, *Geoponica*, p. 20.

22 Boudreaux, ed., *Catalogus Codicum*, 8:3, p. 193.

23 D. Lehoux, *Astronomy, Weather and Calendars in the Ancient World: Parapegmata and Related Texts in Classical and Near-Eastern Societies* (Cambridge: Cambridge University Press, 2007), [hereafter Lehoux, *Astronomy, Weather and Calendars*], pp. 164, 392-99. The text is C. Baroccianus 131, folios 423-423v, in the Bodleian Library, Oxford.

24 S. Weinstock, ed., *Catalogus Codicum Astrologorum Graecorum, Volume 9:1* (Brussels: Lamertin, 1951), pp. 128-37.

25 Lehoux, *Astronomy, Weather and Calendars*, p. 398, n. 204.

Astronomically, in order to keep the calendrical months in line with the seasons of the year, every two to three years an extra month is added, or intercalated, to the year — seven times in a fixed position in the nineteen-year cycle.[26] It is accepted that the nineteen-year cycle was standardised during the Persian period in early fifth century BCE Babylonia,[27] some fifty years before Meton of Athens, whose name is attributed to the cycle.[28] An important feature of the cycle is that the solar and lunar positions return to the same point, and same lunar phase, on or around the same solar date, once in every 235 mean lunar months (nineteen years and seven intercalary months).[29]

It is not certain whether the proto-Jewish calendar — or any of the pre-rabbinical calendars —would have intercalated a second Elul (the sixth month), following the Babylonian practice of adding a second Ululu in the first year of the nineteen-year cycle.[30] Stern argues that they probably did not because that would mean a shift of position of the seventh month when important biblical festivals are held,[31] although it is probable that this

---

26 W. Kendrick Pritchett and O. Neugebauer, *The Calendars of Athens* (Cambridge, Massachusetts: Harvard University Press, 1947), p. 6; O. Neugebauer, *Astronomical Cuneiform Texts* (New York: Springer-Verlag, 1982 [first edition Princeton, N.J-Lund Humphries, 1955]), p. 33; Alan C. Bowen and Bernard R. Goldstein, 'Meton of Athens and Astronomy in the Late Fifth Century BC', in E. Leichty et al., eds., *A Scientific Humanist: Studies in Memory of Abraham Sachs* (Philadelphia: Occasional Publications of the Samuel Noah Kramer Fund, 1988), [hereafter Bowen and Goldstein, 'Meton'], p. 42, n. 17; O. Neugebauer, *The Exact Sciences in Antiquity*, (New York: Dover, 1969), [hereafter, Neugebauer, *Exact Sciences*], p. 7.

27 J. Britton, 'Treatments of Annual Phenomena in Cuneiform Sources', in J.M. Steele and A. Imhausen, eds., *Under One Sky: Astronomy and Mathematics in the Ancient Near East* (Münster-Ugarit-Verlag, 2002), [hereafter Britton, 'Treatments of Annual Phenomena'], pp. 33-36; J. Britton, 'Calendars, Intercalations and Year-Lengths', in John M. Steele, ed., *Calendars and Years: Astronomy and Time in the Ancient Near East* (Oxford: Oxbow Books, 2007), pp. 122-24; J. Britton and C. Walker, 'Astrology and Astronomy in Mesopotamia', in C. Walker, ed., *Astronomy Before the Telescope* (New York: St Martin's Press), pp. 45-46; O. Neugebauer, *History of Ancient Mathematical Astronomy* (New York: Springer-Verlag, 1975), p. 622; F. Rochberg-Halton, 'Calendars: Ancient Near East', in *The Anchor Bible Dictionary, Volume 1* (New York: Doubleday, 1992), pp. 810-11.

28 Bowen and Goldstein, 'Meton', p. 50; B.R. Goldstein, 'A Note on the Metonic Cycle', *Isis*, (1966), Vol. 57(1), pp. 115-16.

29 Neugebauer, *The Exact Sciences,* p. 95; cf. T. Boiy, *Late Achaemenid and Hellenistic Babylon* (Leuven: Peeters, 2004), p. 278 (The Seleucid king Antiochus III attended the New Year festival in Babylon 'on the same day, nineteen years later', in 205/4BCE).

30 Britton, 'Treatments of Annual Phenomena', pp. 33-36 (figs. 3-4); F. Rochberg-Halton, 'Astronomy and Calendars in Ancient Mesopotamia', in J.M. Sasson, ed., *Civilisations of the Ancient Near East, Volume 3* (New York: Schreibner, 1995), [hereafter Rochberg-Halton, 'Astronomy and Calendars'], p. 1938.

31 Stern, *Calendar and Community: A History of the Jewish Calendar 2nd Century BCE-10th Century CE* (Oxford: Oxford University Press, 2001), p. 31; cf. B.Z. Wacholder and D.B.

practice did take place because the rabbis banned it for precisely that reason.[32]

How the 4Q318 zodiac calendar may work and how this information can be extracted from the data in the manuscript will now be illustrated. The lunar year is 354 days long; the solar year is approximately 365¼ days long — 11¼ days longer (the difference between the solar and lunar year is known as the *epact*).[33] The Hebrew calendar as it has evolved today is intercalated according to a nineteen-year cycle of twelve twelve-month years and seven thirteen-month years, so that the festivals will fall in the correct seasons.[34] The 'regular' year is 354 days of alternating six hollow months of 29 days followed by full months of 30 days.[35] If an extra 30-day month were not intercalated every two to three years, the festivals would slip back 11¼ days every year, so that in eight years, the date of the spring festival of Passover, when a lamb is slaughtered, would fall during the winter months, when there are no lambs.

The twelfth month in an intercalary year is doubled: Adar I, the additional month of 30 days, is placed before Adar II, which has 29 days, in keeping with the alternating sequence of full and hollow months. As shall be seen, the accumulative, astronomical effect of the 11¼-day epact over two to three years is reflected in the position of the moon in the zodiac in relation to the calendar.

When reconstructed, it is evident that 4QZodiac Calendar has 360 days,[36] the number of degrees in the zodiac. A 360-day calendar is known from the Ethiopic manuscripts of the *Astronomical Book of Enoch* (also known as the *Book of Luminaries*, *1 Enoch,* chapters 72-82). The textual history of *1 Enoch* is complex and involves conflicting pericope on the calendrical status of four additional days between the seasons, thereby creating a 364-day year (*1 En* 75: 1-2; 82: 4-6, 9-11).[37] Neither the 360-day nor 364-day year-lengths are

---

Weisburg, 'Visibility of the New Moon in Cuneiform and Rabbinic Sources', *Hebrew Union College Annual*, (1971), Vol. 42, [hereafter Wacholder and Weisburg, 'Visibility of the New Moon'], pp. 235-39.

32 Babylonian Talmud, Sanhedrin 12a; see, Wacholder and Weisburg, 'Visibility of the New Moon', p. 237.

33 Bowen and Goldstein, 'Meton', p. 42.

34 J.B. Segal, 'Intercalation and the Hebrew Calendar', *Vetus Testamentum*, (1957), Vol. 7, pp. 263-68.

35 A. Spier, *The Comprehensive Hebrew Calendar: Twentieth to Twenty-Second Century*, (Jerusalem: Feldheim, 1986), pp. 15-16.

36 M. Albani, *Astronomie und Schöpfungsglaube: Untersuchen zum Astronomischen Henochbuch, Volume 68, Wissenschaftliche Mono-graphien zum Alten und Neuen Testament* (Neukirchen-Vluyn: Verlag, 1994), pp. 83-87; Albani, 'Der Zodiakos in 4Q318', pp. 20-21: Albani, 'Horoscopes', p. 300; Greenfield et al., '318. 4QZodiology and Brontology ar', pp. 270-71.

37 J.C. VanderKam, *Calendars in the Dead Sea Scrolls, Measuring Time* (London, Routledge, 1998), pp. 17-27; idem, 'Calendars: Ancient Israelite and Early Jewish', in D.N. Freedman, ed., *Anchor Bible Dictionary, Volume 1,* p. 818; Translations: G.W.E.

extant in the Aramaic fragments of the *Astronomical Book of Enoch*, which are among the oldest scrolls found at Qumran. There are differences as well as significant overlaps with the Ethiopic texts.[38]

The 360-day year is attested in Mesopotamian sources and dates back to an administrative system in the late third millennium BCE.[39] There is a scholarly dispute between Albani, Horowitz and Koch as to whether an ideal, solar-stellar year of 360 days is presupposed, in the MUL. APIN.[40] Brown, like Koch, argues that 360 days is assumed, but not stated, in the text.[41] The year-length is evident in the seventh-century BCE cuneiform text, *The Diviner's Manual,* in which an astronomer seeks permission from the king to intercalate to prevent evil, ostensibly an observatory method of keeping the calendar in check.[42]

In a similar vein, the report of the Assyrian scholar Balasî to the king urges an intercalation to thwart a malefic prediction, although the length of the year is not included in the report:

Nickelsburg and J.C. VanderKam, *1 Enoch: A New Translation* (Minneapolis: Fortress, 2004), [hereafter Nickelsburg and VanderKam, trans., *I Enoch*], pp. 96-116; Matthew Black, trans., *The Book of Enoch or 1 Enoch: A New English Edition with Commentary and Notes, in consultation with J.C. VanderKam, with an Appendix on the 'Astronomical' Chapters (72-82) by Otto Neugebauer, Studia in Veteris Testamenti Pseudepigrapha, Volume 7* (Leiden: Brill, 1985), pp. 386-419; M. Knibb, trans., *The Ethiopic Book of Enoch: A New Edition in the Light of the Aramaic Dead Sea Fragments: In Consultation with Edward Ullendorff, Volume 2* (Oxford: Clarendon, 1978), pp. 167-92.

38 Wise et al., *Dead Sea Scrolls*, pp. 295-303; J.C. VanderKam, 'Sources for the Astronomy in 1 Enoch 72-82', in C. Cohen et al., eds., *Birkat Shalom: Studies in the Bible, Ancient Near Eastern Literature and Postbiblical Judaism Presented to Shalom M. Paul on the Occasion of His Seventieth Birthday* (Winona Lake, Indiana: Eisenbrauns, 2008), pp. 965-78 [hereafter, VanderKam, 'Sources for the Astronomy in 1 Enoch 72-82']; J.T. Milik, *The Books of Enoch: Aramaic Fragments of Cave 4 with the Collaboration of Matthew Black* (Oxford: Clarendon, 1976), [hereafter Milik, *The Books of Enoch*], pp. 7, 274-79; Nickelsburg and VanderKam, trans., *1 Enoch*, pp. 104, 106, 108-9, 114, 116.

39 L. Brack-Bernsen, 'The 360-Day Year in Mesopotamia' in J.M. Steele, ed., *Calendars and Years* (Oxford: Oxbow, 2007), pp. 83-100; R.K. Englund, 'Administrative Time-Keeping in Ancient Mesopotamia', *Journal of the Economic and Social History of the Orient,* (1988), Vol. 31(2), pp. 121-85.

40 Bibliographical details summarised in VanderKam, 'Sources for the Astronomy in 1 Enoch 72-82', pp. 973-78.

41 D. Brown, *Mesopotamian Planetary Astronomy-Astrology* (Groningen: Styx, 2000), [hereafter Brown, *Mesopotamian Planetary Astronomy-Astrology*], pp. 119-20.

42 L. Oppenheim, 'A Babylonian Diviner's Manual', *Journal of Near Eastern Studies,* (1974), Vol. 33(2), p. 200, lines 57-71; C. Williams, 'Signs from the Sky, Signs from the Earth: The Diviner's Manual Revisited', in *Under One Sky* (Münster: Ugarit-Verlag, 2002), pp. 474-75, 482; Brown, *Mesopotamian Planetary Astronomy-Astrology*, p. 211; W. Horowitz, *Mesopotamian Cosmic Geography* (Winona Lake, Indiana: Eisenbrauns, 1998), p. 151.

> Let them intercalate a month: all the stars of the sky have fallen behind. Month XII must not pass unfavourably. Let them intercalate.[43]

Brown argues that the intercalary month would be added principally for divinatory purposes, in order that the stars and planets did not rise on calendrically inauspicious dates.[44]

The most relevant study to 4QZodiac Calendar involving the 360-day year is that of the so-called 'dodekatemoria' and 'Kalendertexte' cuneiform tablets from the late fourth and third century BCE in which the zodiac signs and months are substituted by numbers.[45] With 4Q318, it is arguably possible to see that the calendar was intercalated because the data in the text itself shows that the months are aligned to their correct season by their corresponding lunar zodiac signs. Below (see Fig. 1), is a reconstruction and translation of month twelve, Adar. (The column number 8 is based on a hypothetical reconstruction of the full manuscript commencing with the first month).

The twelfth luni-solar month, Adar, is aligned from February to March, which corresponds approximately to the solar zodiac sign of Pisces. If the ideal conjunction for Adar took place when the sun and moon were at 18° Pisces in a 360-day year, the 4Q318 zodiac calendar would overlap with the solar zodiac.[46]

1	ADAR. On the 1st and on the 2nd, Aries; on the 3rd and on the 4th, Taurus; on the 5[th and on the 6th and on the 7th, Gemini;]
2	on the 8th and on the 9th, C[ancer; on the 10th and on the 11th, L]eo; on the 12th [and on the 13th and on the 14th,]
3	Vir[go]; on the 15th and on [the 16th, Libra; on the 1]7th and on the 1[8th, Scorpio;]
	^ 21st^
4	[On the 1]9th, and on the 20th, Sagitt[arius; on the 22nd and on the 23rd, Cap]ricorn; [on the 24th and on the 25th]
5	Aqu[arius]; on the 26th and on the 2[7th and on the 2]8th, Pi[sces; on the 29th and the 30th,]
6	Arie[s.] vacat [If in Taurus] it thunders ...

Figure 1: 4Q318 col. 8, lines 1-6 (Adar 1-Adar 30).

43 H. Hunger, *Astrological Reports to Assyrian Kings*, State Archives of Assyria 8 (Helsinki: Helsinki University Press, 1992), p. 57, lines 8-10.

44 Brown, *Mesopotamian Planetary Astronomy-Astrology*, pp. 123-24, 195-97.

45 Lis Brack-Bernsen and John M. Steele, 'Babylonian Mathemagics: Two Astronomical-Astrological Texts', in C. Burnett et al., eds., *Studies in the History of the Exact Sciences in Honour of David Pingree* (Leiden: Brill, 2004), [hereafter Brack-Bernsen and Steele, 'Babylonian Mathemagics'), pp. 95-121, table 8, p. 119.

46 Since the ideal conjunction for Nisan in the text is probably 18° Aries.

The full moon moving through the zodiac on day 13-14 would be in the opposite sign to Pisces, which is Virgo, as indeed it is in 4Q318 (see Fig. 1).[47] This author's doctoral research shows that the data in 4Q318 correlate with the moon's position in the zodiac on dates in the Jewish luni-solar calendar following an intercalation. In the Jewish calendar (which is schematic[48]), and the 360-day Mesopotamian zodiacal calendar explicated by Brack-Bernsen and Steele,[49] the first day of each month corresponds to the first lunar crescent and the full moon occurs on days fourteen or fifteen.

Therefore, one can check that the data in 4Q318 is reasonably correct both from the zodiacal position of the moon, and the month-date. If the months and days were not given in the text, then it would be a simple astronomical table, but by including months and days, it becomes a zodiac calendar. To be specific it becomes a schematic lunar ephemeris: a tabulation of the moon-sign according to the day of the month. This finding is significant in the study of the calendars in the Dead Sea Scrolls because the variety of calendrical texts in Hebrew found at Qumran do not have any explicit evidence of intercalation and it is not known how the 364-day priestly calendars could have functioned in practice.[50] For example, Glessmer and VanderKam have postulated different possible solutions, and Beckwith argues that no intercalation took place at all (if so, it is unclear how the harvest festivals and Passover could have been observed).[51]

It will be useful to consider the scientific basis of 4QZodiac Calendar in antiquity. As can be seen from the tabular arrangement in Table 1, below, the zodiacal signs are in their correct order both horizontally and vertically when laid out in a grid. The passage of the moon in the zodiac is given a recurring schematic pattern of two and three days: three days to move through a major lunar phase every seven days, and two days per sign between the other phases. In essence, the moon travels about two-and-a-half days to traverse a sign, but 4Q318 does not deal with fractions.

---

47 The day begins at sunset in Mesopotamia, see Sachs and Hunger, *Astronomical Diaries*, p. 15.

48 Leo Depuydt, 'History of the heleq', in J.M. Steele and A. Imhausen, eds., *Under One Sky: Astronomy and Mathematics in the Ancient Near East* (Münster: Ugarit-Verklag, 2002), pp. 83-84.

49 Brack-Bernsen and Steele, 'Babylonian Mathemagics', p. 119, table 8.

50 R.T. Beckwith, *Calendar and Chronology, Jewish and Christian: Biblical, Intertestamental and Patristic Studies* (Leiden: Brill, 2001), [hereafter Beckwith, *Calendar and Chronology*], pp. 125-40; J. Ben-Dov, *Head of All Years: Astronomy and Calendars at Qumran in their Ancient Context, Studies on the Texts of the Desert of Judah, Volume 78* (Leiden: Brill, 2008), pp. 18-20.

51 U. Glessmer, 'Calendars in the Qumran Scrolls', in J.C. Vanderkam and P.W. Flint, eds., *The Dead Sea Scrolls After Fifty Years* (Leiden: Brill, 1998), pp. 263-68; VanderKam, *Calendars*, pp. 80-84; Beckwith, *Calendar and Chronology*, pp. 139-40.

	Nisan	Iyyar	Sivan	Tammuz	Av	Elul	Tishri	**Heshvan**	Kislev	Tevet	Shevat	Adar
1	♉	♊	♋	♌	♍	♎	♏	♐	♑	♒	♓	♈
2	♉	♊	♋	♌	♍	♎	♏	♐	♑	♒	♓	♈
3	♊	♋	♌	♍	♎	♏	♐	♑	♒	♓	♈	♉
4	♊	♋	♌	♍	♎	♏	♐	♑	♒	♓	♈	♉
5	♋	♌	♍	♎	♏	♐	♑	♒	♓	♈	♉	♊
6	♋	♌	♍	♎	♏	♐	♑	♒	♓	♈	♉	♊
7	♋	♌	♍	♎	♏	♐	♑	♒	♓	♈	♉	♊
8	♌	♍	♎	♏	♐	♑	♒	♓	♈	♉	♊	♋
9	♌	♍	♎	♏	♐	♑	♒	♓	♈	♉	♊	♋
10	♍	♎	♏	♐	♑	♒	♓	♈	♉	♊	♋	♌
11	♍	♎	♏	♐	♑	♒	♓	♈	♉	♊	♋	♌
12	♎	♏	♐	♑	♒	♓	♈	♉	♊	♋	♌	♍
13	♎	♏	♐	♑	♒	♓	♈	♉	♊	♋	♌	♍
14	♎	♏	♐	♑	♒	♓	♈	♉	♊	♋	♌	♍
15	♏	♐	♑	♒	♓	♈	♉	♊	♋	♌	♍	♎
16	♏	♐	♑	♒	♓	♈	♉	♊	♋	♌	♍	♎
17	♐	♑	♒	♓	♈	♉	♊	♋	♌	♍	♎	♏
18	♐	♑	♒	♓	♈	♉	♊	♋	♌	♍	♎	♏
19	♑	♒	♓	♈	♉	♊	♋	♌	♍	♎	♏	♐
20	♑	♒	♓	♈	♉	♊	♋	♌	♍	♎	♏	♐
21	♑	♒	♓	♈	♉	♊	♋	♌	♍	♎	♏	♐
22	♒	♓	♈	♉	♊	♋	♌	♍	♎	♏	♐	♑
23	♒	♓	♈	♉	♊	♋	♌	♍	♎	♏	♐	♑
24	♓	♈	♉	♊	♋	♌	♍	♎	♏	♐	♑	♒
25	♓	♈	♉	♊	♋	♌	♍	♎	♏	♐	♑	♒
26	♈	♉	♊	♋	♌	♍	♎	♏	♐	♑	♒	♓
27	♈	♉	♊	♋	♌	♍	♎	♏	♐	♑	♒	♓
28	♈	♉	♊	♋	♌	♍	♎	♏	♐	♑	♒	♓
29	♉	♊	♋	♌	♍	♎	♏	♐	♑	♒	♓	♈
30	♉	♊	♋	♌	♍	♎	♏	♐	♑	♒	♓	♈

Table 1: 4Q318 with zodiac glyphs.
Key to symbols: Aries ♈; Taurus ♉; Gemini ♊; Cancer ♋; Leo ♌; Virgo ♍; Libra ♎; Scorpio ♏; Sagittarius ♐; Capricorn ♑; Aquarius ♒; Pisces ♓.

The calendar is divided into 12 synodic months (from one lunar phase to the next) of 30 days and it begins with the first month, Nisan, in the spring. In a month, the moon traverses the 12 signs of the zodiac plus the first sign that it travelled through to catch up with the sun, which will have moved one zodiac sign ahead during that month. Hence, in each 30-day month in the 360-day calendar, the moon will have travelled through 13 signs. In contrast, the sun takes one month of 30 days to traverse each sign and a year of 360 days to travel through all 12 signs (not explicit in the manuscript).

Testing the zodiac calendar

When the days of the months in 4QZodiac Calendar are converted to Julian or Gregorian dates[52] and the zodiac sign of the moon is computed accordingly, the results yielded are significant. Fig. 2 is a sample of data for 14 Adar, converted to the Gregorian calendar for the contemporary period. Due to the 11¼ day epact, there is an incremental eleven-day difference each year between the Hebrew (lunar) date and the Gregorian (solar) calendar, which is remedied by the two or three year intercalary month.[53] As stated above, the two luminaries return to the same calendrical position after nineteen years.

14 ADAR	LUNAR SIGN	PHASE
14 March 2006	Virgo	Full
4 March 2007	Virgo	Full
20 February 2008*	Leo	Full
21 March 2008	Virgo	Full
10 March 2009	Virgo	Full
28 February 2010	Virgo	Full

Figure 2: 4Q Zodiac Calendar (14 Adar converted to the Gregorian calendar, 2006 to 2010 at noon).
*Intercalary Year, Adar I.

As can be seen from the converted dates, the full moon is in Virgo and the sun is in Pisces, except in Adar I when, due to the epact, the months are three weeks behind the season. Fuller data for the calendar reveals a similar pattern in the years requiring a correction when, 'the stars have fallen

52 Calendar converter program: *Kaluach 3*, online at: http://www.fourmilab.ch/documents/calendar/ [accessed 15 July 2009]. Note, in the Hebrew calendar the day begins and ends at sunset; in this programme, it commences at 12 noon.

53 Cf. Rochberg, 'Astronomy and Calendars', p. 1931.

behind', in particular in Adar I, when an intercalation is due.[54] The diviner using 4Q318 might calibrate the moon's position for every third year or so, when the moon will be in a slightly earlier zodiacal position.

Perhaps the complete scroll contained the variations for non-intercalary years; if so, the manuscript, including the brontologion, would be about 3.79 metres long. Rolled up, it would be bulky to carry, but possible; if it was a template for one year (the practitioner used his knowledge to calculate the moon's position in non-intercalary years), the scroll would be about 1.38 metres long and easier to transport.[55] However, no extant fragments for other years have been identified.

In order for there to be a correlation using contemporary Jewish calendar conversion tables, it is possible that the Qumran zodiac calendar was structured to similar synodic cycles to those of the rabbinical calendar known today (that is, from the ideal first crescent). The early development and the history of the adoption of the Jewish calendar prior to the tenth century CE, however, is uncertain and remains an area of scholarly research.[56]

On a similar note, the tropical zodiac, defined as beginning at 0° Aries longitude, appears to have been extant during the period that this scroll was copied or composed. It is used in 4QZodiac Calendar; if it were not, the position of the moon in the zodiac in modern ephemerides would not correspond with 4Q318. This norm is attributed to the Greeks[57] and has been preserved in astrological tables by western astrologers. The antiquity of the zero-degree cardinal point has been recently archaeologically confirmed in the Antikythera Mechanism (150-100BCE) on the front, 'zodiac' dial, where the letter *alpha* is visible next to 0° Libra.[58]

Whatever its length, the Aramaic zodiac calendar simply required a rudimentary understanding of astronomy, and a knowledge of which years were intercalary, in order to be read. As we shall now see, calendrical education, particularly with regard to the zodiac, was regarded as elementary for the literate within the wider society.

---

54 Discussed in this author's doctoral thesis.

55 I thank Bernard Eccles for his feedback on this point.

56 Stern, *Calendar and Community*, pp. 175-81; Wacholder and Weisburg, 'Visibility of the New Moon', p. 239.

57 James Evans, *The History and Practice of Ancient Astronomy* (Oxford: Oxford University Press, 1998), pp. 213-14.

58 Derek J. De Solla Price, *Gears from the Greeks: The Antikythera Mechanism — A calendar computer from ca. 80 BC: Transactions of the American Philosophical Society, 64:7* (Philadelphia: American Philosophical Society, 1974), p. 18; Robert Hannah, *Time in Antiquity* (London, Routledge, 2008), pp. 48-49, note 59. See Polynomial Texture Mapping (PTM) image AK31a, online at: http://www.hpl.hp.com/research/ptm/antikythera_mechanism/full_resolution_ptm.htm?jumpid=reg_R1002_USEN [accessed 10 January 2009].

Cultural background

Zodiac calendars began to appear in the Greek, Ptolemaic and Mesopotamian worlds in the fifth, fourth and third centuries BCE, mainly independently of corresponding omen texts. Some of these calendars, which exist in archaeological artefacts such as the Antikythera Mechanism, parapegmata and papyri, are intricate mathematical constructs which coordinate the sun, moon and the stars. Great efforts appear to have been made to impart this science in the late antique world in literary and scientific texts and three-dimensional artefacts.

Vitruvius, the Roman, writer, architect and engineer, explained the astrological, cosmological, calendrical and scientific importance of the solar and lunar zodiac in many places throughout Book 9 of *On Architecture*, which he presented to the Emperor Augustus in the mid-20s BCE:

> 9.5. These signs, therefore, are twelve in number and each individual sign occupies one-twelfth part of the firmament, and all of them are constantly rotated from east to west ... the moon, traversing its circuit in a little more, by about an hour ... completes a lunar month by returning to the sign in which it had first set out
>
> 6. In the turning of a month, the sun ... traverses the space of a single sign, that is, one-twelfth part of the firmament. By travelling across the distance of twelve signs in twelve months, it completes the interval of the revolving year then it returns to the sign in which it began. In other words, that [zodiacal] circuit which the moon runs thirteen times in twelve months, the sun measures out only once in the same number of months.[59]

Ovid, in imparting the history of Roman calendar, writes in *Fasti*, (early first century CE), that during the prehistoric Roman period of Romulus (eighth century BCE) the year had 10 months.[60] Then, he states, people were ignorant about the stars [no reasons are given], and the passages of the sun and the moon were not synchronised with the zodiac calendar:

---

59 Vitruvius, *Ten Books on Architecture,* trans. I.D. Rowland (Cambridge: Cambridge University Press, 1999), p. 110.

60 Bonnie Blackburn and Leofranc Holford-Strevens, *The Oxford Companion to the Year: An Exploration of Calendar Customs and Time-reckoning* (Oxford: Oxford University Press, 1999), p. 669; E.G. Richards, *Mapping Time: The Calendar and its History* (Oxford: Oxford University Press, 2000), p. 207; Denis Feeny, *Caesar's Calendar: Ancient Time and the Beginnings of History* (Berkeley: University of California Press, 2008), pp. 202-3; S.J. Green, *Ovid. Fasti I: A Commentary, Mnemosyne Supplementa 251* (Leiden: Brill, 2004), pp. 44-47; A.E. Samuel, *Greek and Roman Chronology* (Munich: Oscar Beck, 1972), [hereafter Samuel, *Greek and Roman Chronology*], pp. 164-65, 167-70; Robert Hannah, *Greek and Roman Calendars: Constructions of Time in the Classical World* (London: Duckworth, 2005), p. 99.

> ... Who had then noticed ... that the [zodiac] signs which the brother [Apollo, the sun] travels through in a long year, the horses of the sister [Diana, the moon] traverse in a single month?[61]

With regards to Jewish writings, it would appear that Philo (c. 20BCE-50CE) knew of a lunar and solar zodiac to which he refers in several of his works. For example, he explicates the periodic relationship in a commentary relating to Genesis 37: 9-11, Joseph's second dream that the sun, moon and 11 stars bowed down to him:

> The sun and moon they say, ever revolve along the circle [*zodia*] and pass through each of the signs, though the two do not move at the same speed, but at unequal rates as measured in numbers, the sun taking thirty days and the moon about a twelfth of that time, that is, two and half days.[62]

Philo's zodiacal references are extensive and he and Josephus after him used the zodiac in prose to explain both scientific thought and cosmology in the description of the vestments of the High Priest and the Temple. The symbolic references to the years (sun) and to the months (moon) are kept separate by both writers.[63] Philo refers to the lunar zodiac in his explanation of the *noumenia*, the new moon festival:

> ... the moon traverses the zodiac in a shorter fixed period than any other heavenly body. For it accomplishes that revolution in the span of a single month, and therefore, the conclusion of its circuit, when the moon ends its course at the starting point at which it began ....[64]

Returning to the cosmological passages in the Dead Sea Scrolls, references to the zodiac, the luminaries and birth-times are attested.[65] Milik drew a

---

61 Ovid, *Fasti.* 3.97-10, trans. Sir James George Grazer (Cambridge, Massachusetts: Harvard University Press, 1959), pp. 126-29.

62 Philo, *On Dreams 2, Volume 5*, trans. F.H. Colson and G.H Whitaker (London: William Heinemann [Cambridge, Mass.: Harvard University Press], 1934), pp. 492-93.

63 See Helen R. Jacobus, '4Q318: A Jewish zodiac calendar at Qumran', in C. Hempel, ed., *The Dead Sea Scrolls: Texts and Context* (Leiden: Brill, 2010).

64 Philo, *On Special Laws 2, Volume 7*, trans. F.H. Colson (London: William Heinemann [Cambridge, Massachusetts: Harvard University Press], 1937), pp. 392-93.

65 M. Morgenstern, 'The Meaning of *Beit Moladim*', *Journal of Jewish Studies*, (2000), Vol. 51(1), pp. 141-44; E.J.C. Tigchelaar, 'Your Wisdom and Your Folly: The Case of 1-4QMysteries', in F. García Martínez, ed., *Wisdom and Apocalypticism in the Dead Sea Scrolls and in Biblical Tradition* (Leuven: Peeters, 2003), p. 88; Francis Schmidt, 'Recherche son thème de geniture dans le mystère de ce qui doit être': Astrologie et prédestination à Qoumrân', in A. Lemaire and S.C. Mimouni, eds., *Qoumran et le Judaïsme du Tourant de Notre Ère: Collection de la Revue des Ètudes Juives, Volume 40* (Leuven: Peeters, 2006), pp. 57-8; M. Goff, 'The Mystery of Creation in 4QInstruction',

connection between the zodiac in 4Q318 and a hymn in the *Community Rule* (1QS col 10, lines 1-5), a document which is clearly regarded as sectarian. Milik translated the pericope as follows:

> When the lights shine forth from the Holy Dwelling-Place, and when also they retire (lit. are gathered) to the Place of Glory, when the constellations (of the Zodiac) make (their), entrance on the days of the new moon, and their circuit at their positions every new moon succeeding one after another, it is a Great Day for the Holy of Holies, and a sign for the unlocking of everlasting mercies corresponding to the beginning of the constellations (of the Zodiac), to last for all time to come.[66]

However, he is the only scholar to have translated this text in terms of references to the zodiac.[67] The sublime, cosmological poems from the Thanksgiving Hymns (Hebrew: *Hodayot*) in the Dead Sea Scrolls are religious texts also containing philosophical ideas about astronomy, pre-destination and eternity. The Thanksgiving Scroll (col. 20, lines 7-14), describes the orbits of the sun and moon, the renewal of each day and the cycle of seasons and festivals. The author connects the circuits of the luminaries — pre-set forever from the beginning of time — to God. The use of a sacred calendar, in which the festivals are fixed in the right order by their 'signs', is ambiguous and could be interpreted as referring to the zodiac. The poem describes the orbits of all the heavenly bodies as predictable as is everything on earth, due to a divine and predetermined plan:

> 7. [For the instruct]or, [th]anksgiving and prayer for prostrating oneself and supplicating continually at all times: with the coming of light.
> 8. for [its] domin[ion]; the midpoints of the day with respect to its arrangement according to the rules of the great light; when it turns to evening and light goes forth
> 9. at the beginning of the dominion of darkness at the time appointed for night; at its midpoint, when it turns toward morning; and at the time that
> 10. it is gathered in to its dwelling place before (the approach of) light, at the departure of night and the coming of day, continually, at all the

Dead Sea Discoveries, (2003), Vol. 10(2), p. 168; Wise et al., *Dead Sea Scrolls*, p. 111; L.H. Schiffman, '4Q299 (4QMysteriesa)' in D.W. Parry and E. Tov, eds., *The Dead Sea Scrolls Reader, Part 4: Calendrical and Sapiential Texts* (Leiden: Brill, 2004), p. 203.

66 J.T. Milik, *The Books of Enoch*, p. 187.

67 Cf. E. Qimron and J.H. Charlesworth, trans., 'Rule of the Community and Related Documents', in J.H. Charlesworth, ed., *The Dead Sea Scrolls: Hebrew, Aramaic, and Greek Texts with English Translations, The Princeton Theological Seminary Dead Sea Scrolls Project 1* (Tübingen: Mohr-Siebeck, 1994), p. 43; P.S. Alexander and G. Vermes, trans., *Discoveries in the Judean Desert, Volume 26*, pp. 60, 117-18, 122; Wise et al., *Dead Sea Scrolls*, p. 132.

> 11. birthings of time, the foundations of the seasons, and the cycle of the festivals in the order fixed by their signs, for all
> 12. their dominion in proper order, reliably, at the command of God. It is a testimony of that which exists. This is what shall be,
> 13. and there shall be no end. Apart from it nothing has existed nor shall yet be. For the God of knowledge
> 14. has established it, and there is none other with him.[68]

In the following extracts from the Thanksgiving Scroll (col 9, lines 9-15, 20-22, 25-26), angels, stars, shooting stars, lightning, the destiny of future generations and inscribed heavenly tablets are included in the sacred, cosmological design.

> 9. And in your wisdom [ ] eternity, and before you created them, you know {all} their deeds
> 10. for everlasting ages. And [without you no]thing is done and nothing is known without your will. You formed
> 11. every spirit, and [their] work [you determin]ed, and the judgement for all their deeds. You yourself stretched out the heavens
> 12. for your glory, and all [ ] you [de]termined according to your will, and powerful spirits according to their laws, before
> 13. they came to be ho[ly] angels [and]*m* eternal spirit in their dominions: luminaries according to their mysteries,
> 14. stars according to [their] paths, [stor]m [winds] according to their task, shooting stars and lightning according to their service, and storehouses
> 15. devised for th[eir] purposes [ ] according to their mysteries. ....
> 20. And you allotted it to all their offspring according to the number of generations of eternity
> 21. and for all the everlasting years ... And in the wisdom of your knowledge you determ[i]ned their des[t]iny before
> 22. they existed. According to your wi[ll] everything [comes] to pass; and without you nothing is done. *vacat*
> 25-26. ... Everything is engraved before you in an inscription of record for all everlasting seasons and the numbered cycles of the eternal years with their appointed times.[69]

4Q318 is culturally eclectic and cannot be neatly categorised; yet it can be shown empirically that, unlike the Hebrew calendars of the priestly courses

---

68 Carol Newsom, trans., *Qumran Cave 1: III: 1QHodayota with Incorporation of 4QHodayota-f and 1QHodayotb: Discoveries in the Judaean Desert, Volume 40* (Oxford: Clarendon, 2008), p. 259.

69 Carol Newsom, trans., *Qumran Cave 1: III: 1QHodayota with Incorporation of 4QHodayota-f and 1QHodayotb: Discoveries in the Judaean Desert, Volume 40* (Oxford: Clarendon, 2008), p. 130; Carol Newsom, *The Self as Symbolic Space: Constructing Identity and Community at Qumran*, pp. 222-23, 226.

at Qumran, it has a definite astronomical basis. There is textual support in the Dead Sea Scrolls to suggest that the cosmological perspective and scientific basis of such a paradigm may have been attributed a religious and spiritual significance. Liturgical themes in the Dead Sea Scrolls interweaving concepts of repeated astronomical cycles with timelessness and eternity could be inspired by a zodiacal calendar, which integrates the sun, moon and the stars: a biblical function of the luminaries described in Genesis 1: 14-19.

There is also evidence from across the Hellenistic world to suggest that zodiac calendars existed in different cultural forms in the wider society. Its presence in Qumran Cave 4 where a variety of non-zodiacal Hebrew calendars were found implies that there was a wide range of calendrical activity at the turn of the era, as indeed there was throughout the region. If one considers that the Julian calendar was introduced in 46/45BCE, gradually displacing most of the lunar calendars except the lunar calendar of the eastern provinces,[70] that should be illustrative of the preoccupation of the period. There is no reason why Jewish groups should not have similarly shared an intellectual interest in the subject.

The 4Q318 zodiac calendar, it is suggested, is the only calendar of this type for which there is an extant primary source that survived. It is the sole such calendar in the Dead Sea Scrolls to be explicitly connected to prognostication. From it we learn that the zodiac calendar could be used as an ephemeris. Historically, the Jewish calendar is based on the Babylonian calendar, which in its earlier form was connected with omen divination, probably as a means of keeping the months synchronised with the stars. What is surprising is that 4QZodiac Calendar is a perpetual calendar that can still be used today, albeit because western astrology is anachronistic. As a result of the preservation and practise of ancient astrology until the present, 4Q318 demonstrates that the tropical zodiac was in use in some 150 years before Ptolemy. Not only may 4Q318 lead us to revisit the historical development and adoption of the pre-rabbinical Jewish calendar, but it might also impact on our knowledge of the history of cosmology, and the practice of astrology amongst monotheistic groups 2,000 years ago.

Bibliography

Abegg, Martin, Peter Flint and Eugene Ulrich, eds., *The Dead Sea Scrolls Bible: The Oldest Known Bible Translated for the First Time into English* (New York: HarperCollins, 1999).

Albani, Matthias, 'Der Zodiakos in 4Q318 und die Henoch-Astronomie', *Mitteilungen und Beiträge der Forschungsstelle Judentum der Theologischen Fakultät Leipzig*, (1993), Vol. 7, pp. 3-42.

70 Samuel, *Greek and Roman Chronology*, pp. 186-88.

Albani, Matthias, *Astronomie und Schöpfungsglaube: Untersuchen zum Astronomischen Henochbuch, Volume 68, Wissenschaftliche Mono-graphien zum Alten und Neuen Testament* (Neukirchen-Vluyn: Verlag, 1994).

Albani, Matthias, 'Horoscopes in the Qumran Scrolls', in P.W. Flint and J.C. VanderKam, eds., *The Dead Sea Scrolls After Fifty Years* (Leiden: Brill, 1999), pp. 279-330.

Albani, Matthias, 'Horoscopes', in vol. 1 of L.H Schiffman and J.C. VanderKam, eds., *Encyclopedia of the Dead Sea Scrolls* (Oxford: Oxford University Press, 2000), pp. 370-73.

Alexander, P.S. and G. Vermes, *Qumran Cave 4: XIX. Serekh ha-Yahad and Two Related Texts*, Volume 26 of *Discoveries in the Judean Desert* (Oxford: Clarendon, 1998).

Beckh, H., ed., *Geoponica sive Cassiani Bassi Scholastici de re Rustica Eclogae* (Leipzig: Biblioteca Teubneriana, 1895).

Beckwith, R.T., *Calendar and Chronology, Jewish and Christian: Biblical, Intertestamental and Patristic Studies* (Leiden: Brill, 2001).

Ben-Dov, J., *Head of All Years: Astronomy and Calendars at Qumran in their Ancient Context, Studies on the Texts of the Desert of Judah, Volume 78* (Leiden: Brill, 2008).

Bickerman, E.J., 'Calendars and Chronology', in W.D. Davies and L. Finkelstein, eds., *The Cambridge History of Judaism, Volume 1* (Cambridge: Cambridge University Press, 1984), pp. 60-69.

Black, Matthew, trans., *The Book of Enoch or 1 Enoch: A New English Edition with Commentary and Notes, in consultation with J.C. VanderKam, with an Appendix on the 'Astronomical' Chapters (72-82) by Otto Neugebauer, Studia in Veteris Testamenti Pseudepigrapha, Volume 7* (Leiden: Brill, 1985).

Blackburn, Bonnie and Leofranc Holford-Strevens, *The Oxford Companion to the Year: An Exploration of Calendar Customs and Time-reckoning* (Oxford: Oxford University Press, 1999).

Brack-Bernsen, Lis and John M. Steele., 'Babylonian Mathemagics: Two Astronomical-Astrological Texts', in C. Burnett et al., eds., *Studies in the History of the Exact Sciences in Honour of David Pingree* (Leiden: Brill, 2004), pp. 95-121.

Brack-Bernsen, L., 'The 360-Day Year in Mesopotamia', in J.M. Steele, ed., *Calendars and Years: Astronomy and Time in the Ancient Near East* (Oxford: Oxbow, 2007), pp. 83-100.

Britton, J., 'Treatments of Annual Phenomena in Cuneiform Sources', in J.M. Steele and A. Imhausen, eds., *Under One Sky: Astronomy and Mathematics in the Ancient Near East* (Münster-Ugarit-Verlag, 2002), pp. 21-78.

Britton, J., 'Calendars, Intercalations and Year-Lengths', in J. M. Steele, ed., *Calendars and Years: Astronomy and Time in the Ancient Near East* (Oxford: Oxbow Books, 2007), pp. 115-32.

Britton, J. and C. Walker, 'Astrology and Astronomy in Mesopotamia', in C. Walker, ed., *Astronomy Before the Telescope* (New York: St Martin's Press), pp. 42-67.

Brown, D., *Mesopotamian Planetary Astronomy-Astrology* (Groningen: Styx, 2000).

Boiy, T., *Late Achaemenid and Hellenistic Babylon* (Leuven: Peeters, 2004).

Boudreaux, P., ed., *Catalogus Codicum Astrologorum Graecorum*, Volume 8:3 (Brussels: Lamertin, 1912).

Bowen, Alan C. and Bernard R. Goldstein, 'Meton of Athens and Astronomy in the Late Fifth Century BC', in E. Leichty et al., eds., *A Scientific Humanist: Studies in Memory of Abraham Sachs* (Philadelphia: Occasional Publications of the Samuel Noah Kramer Fund, 1988), pp. 39-82.

Cohen, Mark. E., *Cultic Calendars of the Ancient Near East* (Bethesda, Maryland: CDL Press, 1993).

Cook, Edward, 'A Divination Text (Brontologion)', in Michael Wise, Martin Abegg and Edward Cook, eds., *The Dead Sea Scrolls: A New Translation* (New York: HarperCollins, 2005).

Davies, Philip R., 'Sect Formation in Early Judaism', in David J. Chalcraft, ed., *Sectarianism in Early Judaism: Sociological Advances* (London: Equinox, 2007), pp. 133-155.

Depuydt, Leo, 'History of the *Heleq*', in J.M. Steele and A. Imhausen, eds., *Under One Sky: Astronomy and Mathematics in the Ancient Near East* (Münster: Ugarit-Verklag, 2002), pp. 79-108.

Englund, R.K., 'Administrative Time-Keeping in Ancient Mesopotamia', *Journal of the Economic and Social History of the Orient,* (1988), Vol. 31(2), pp. 121-85.

Evans, James, *The History and Practice of Ancient Astronomy* (Oxford: Oxford University Press, 1998).

Feeny, Denis, *Caesar's Calendar: Ancient Time and the Beginnings of History* (Berkeley: University of California Press, 2008).

Glessmer, Uwe, 'Calendars in the Qumran Scrolls', in P.W. Flint and J.C. VanderKam, eds., *The Dead Sea Scrolls After Fifty Years,* 2 Volumes (Leiden: Brill, 1998).

Goff, M., 'The Mystery of Creation in 4QInstruction', *Dead Sea Discoveries,* (2003), Vol. 10(2), pp. 163-86.

Goldstein, B.R., 'A Note on the Metonic Cycle', *Isis,* (1966), Vol. 57(1), pp. 115-16.

Green, S.J., *Ovid. Fasti I: A Commentary, Mnemosyne Supplementa 251* (Leiden: Brill, 2004).

Greenfield, J.C. and M. Sokoloff, with A. Yardeni and D Pingree, '318: 4QZodiology and Brontology ar', in Pfann, P., et al., eds., *Discoveries in the Judaean Desert, Volume 36, Qumran Cave 4: XXVI* (Oxford: Clarendon Press, 2000), pp. 259-74.

Hannah, Robert, *Time in Antiquity* (London, Routledge, 2008).

Hannah, Robert, *Greek and Roman Calendars: Constructions of Time in the Classical World* (London: Duckworth, 2005).

Hunger, H., *Astrological Reports to Assyrian Kings,* State Archives of Assyria 8 (Helsinki: Helsinki University Press, 1992).

Horowitz, W., *Mesopotamian Cosmic Geography* (Winona Lake, Indiana: Eisenbrauns, 1998).

Jacobus, H.R., '4Q318: A Jewish Zodiac Calendar at Qumran?', in C. Hempel, ed., *The Dead Sea Scrolls: Texts and Context, Studies on the Texts of the Desert of Judah, Volume 90* (Leiden: Brill, 2010).

Knibb, M., trans., *The Ethiopic Book of Enoch: A New Edition in the Light of the Aramaic Dead Sea Fragments: In Consultation with Edward Ullendorff,* 2 Volumes (Oxford: Clarendon, 1978).

Kollerstrom, Nick, 'Decoding the Antikythera Mechanism', *Astronomy Now,* (2007), Vol. 21(3), pp. 28-31.

Lange, Armin, 'Wisdom and Predestination in the Dead Sea Scrolls', *Dead Sea Discoveries*, (1995), Vol. 2(23), pp. 340-54.

Lange, Armin, 'The Essene Position on Magic and Divination', in M. Bernstein et al., eds., *Legal Texts and Legal Issues: Proceedings of the Second Meeting of the International Organisation for Qumran Studies Cambridge 1995: Published in Honour of Joseph Baumgarten*, STDJ 23 (Leiden: Brill, 1997), pp. 377-435.

Lehoux, Daryn, *Astronomy, Weather and Calendars in the Ancient World: Parapegmata and Related Texts in Classical and Near-Eastern Societies* (Cambridge: Cambridge University Press, 2007).

Nickelsburg G.W.E and J.C. VanderKam, trans., *1 Enoch: A New Translation* (Minneapolis: Fortress, 2004).

Malzbender, Tom, Dan Gelb and the Antikythera Mechanism Research Project, 'Full Resolution PTM Downloads of the Antikythera Mechanism', http://www.hpl.hp.com/research/ptm/antikythera_mechanism/full_resolution_ptm.htm?jumpid=reg_R1002_USEN [accessed 10 January 2010].

Michelson, N.F. and R. Pottinger, *The American Ephemeris for the 21st Century: 2000 to 2050 at Noon* (San Diego: ACS Publications, 1996).

Milik, J.T., *Ten Years of Discovery in the Wilderness of Judea: Studies in Biblical Theology, Volume 26*, trans. J. Strugnell (London: SCM, 1959).

Milik, J.T., *The Books of Enoch: Aramaic Fragments of Cave 4 with the Collaboration of Matthew Black* (Oxford: Clarendon, 1976).

Morgenstern, M., 'The Meaning of *Beit Moladim*', *Journal of Jewish Studies*, (2000), Vol. 51(1), pp. 141-44.

Neugebauer, O., *The Exact Sciences in Antiquity*, second edition (New York: Dover, 1969).

Neugebauer, O., *History of Ancient Mathematical Astronomy*, 3 Volumes (New York: Springer-Verlag, 1975).

Neugebauer, O., *Astronomical Cuneiform Texts* (New York: Springer-Verlag, 1982 [first edition Princeton, N.J-Lund Humphries, 1955]).

Newsom, Carol, Hartmann Stegemann, and Eileen Schuller, *Qumran Cave 1: III: 1QHodayot[a] with Incorporation of 4QHodayot[a-f] and 1QHodayot[b:] Discoveries in the Judaean Desert, Volume 40* (Oxford: Clarendon, 2008).

Newsom, Carol, A., *The Self As Symbolic Space: Constructing Identity and Community at Qumran, Studies on the Texts of the Deserts of Judah, Volume 52* (Leiden: Brill, 2004).

Oppenheim, L., 'A Babylonian Diviner's Manual', *Journal of Near Eastern Studies*, (1974), Vol. 33(2), pp. 197-220.

Ovid, *Fasti*, trans., Sir James George Grazer (Cambridge, Massachusetts: Harvard University Press, 1959).

Owen, Thomas, trans., *Geoponica: Agricultural Pursuits* (London: Spilsbury, 1806).

Philo, *On Dreams 2, Volume 5*, trans. F.H. Colson and G.H Whitaker (London: William Heinemann [Cambridge, Massachusetts: Harvard University Press], 1934).

Philo, *On Special Laws 2, Volume 7*, trans. F.H. Colson (London: William Heinemann [Cambridge, Massachusetts: Harvard University Press], 1937).

Popović, Mladen, *Reading the Human Body: Physionomics and Astrology in the Dead Sea Scrolls and Hellenisitic-Early Roman Period Judaism, Studies on the Texts of the Deserts of Judah, Volume 67* (Leiden: Brill, 2007).

Popović, Mladen, 'Reading the Human Body and Writing in Code: Physionomic Divination and Astrology in the Dead Sea Scrolls', Hilhorst, Anthony, Émile Puech and Eibert Tigchelaar, eds., *Flores Florentino: Dead Sea Scrolls and Other Early Jewish Studies in Honour of Florentino García Martínez: Supplements to the Journal for the Study of Judaism, Volume 122* (Leiden: Brill, 2007), pp. 271-84.

Porten, B., 'The Calendar of Aramaic Texts from Achaemenid and Ptolemaic Egypt', *Irano-Judaica II*, (1990), pp. 13-32.

Price, Derek J. De Solla., *Gears from the Greeks: The Antikythera Mechanism — A calendar computer from ca. 80 BC: Transactions of the American Philosophical Society, 64:7* (Philadelphia: American Philosophical Society, 1974).

Pritchett, Kendrick, W. and O. Neugebauer, *The Calendars of Athens* (Cambridge, Massachusetts: Harvard University Press, 1947).

Qimron, E. and J.H. Charlesworth, trans., 'Rule of the Community and Related Documents', in J.H. Charlesworth, ed., *The Dead Sea Scrolls: Hebrew, Aramaic, and Greek Texts with English Translations, The Princeton Theological Seminary Dead Sea Scrolls Project 1* (Tübingen: Mohr-Siebeck, 1994).

Richards, E.G., *Mapping Time: The Calendar and its History* (Oxford: Oxford University Press, 2000).

Rochberg, Francesca, *The Heavenly Writing: Divination, Horoscopy, and Astronomy in Mesopotamian Culture* (Cambridge: Cambridge University Press, 2004).

Rochberg-Halton, F., 'Calendars: Ancient Near East', in D.N. Freedman, ed., *The Anchor Bible Dictionary, Volume 1* (New York: Doubleday, 1992), pp. 810-14.

Rochberg-Halton, F., 'Astronomy and Calendars in Ancient Mesopotamia', in J.M. Sasson, ed., *Civilisations of the Ancient Near East, Volume 3* (New York: Schreibner, 1995), pp. 1925-40.

Sachs, Abraham J. and Hermann Hunger, *Astronomical Diaries and Related Texts from Babylonia, Volume 1: Diaries from 652 BC to 262 BC* (Vienna: Verlag der Österreichischen Akademie der Wissenschaften, 1988).

Samuel, A.E., *Greek and Roman Chronology* (Munich: Oscar Beck, 1972).

Schiffman, L.H., '4Q299 (4QMysteries[a])', in D.W. Parry and E. Tov, eds., *The Dead Sea Scrolls Reader, Part 4: Calendrical and Sapiential Texts* (Leiden: Brill, 2004), p. 203.

Schmidt, Francis, 'Recherche son Thème de Géniture dans le Mystère de ce Qui Doit Être, Astrologie et Prédestination à Qoumrân', in A. Lemaire and S.C. Mimouni, eds., *Qoumrân et le Judaïsme du Tourant de Notre Ère: Collection de la Revue des Ètudes Juives, Volume 40* (Leuven: Peeters, 2006), pp. 51-62.

Segal, J.B., 'Intercalation and the Hebrew Calendar', *Vetus Testamentum*, (1957), Vol. 7, pp. 250-307.

Spier, A., *The Comprehensive Hebrew Calendar Twentieth to Twenty-Second Century* (Jerusalem: Feldheim, 1986).

Stern, Sacha, *Calendar and Community: A History of the Jewish Calendar 2nd Century BCE-10th Century CE* (Oxford: Oxford University Press, 2001).

Tigchelaar, E.J.C., 'Your Wisdom and Your Folly: The Case of 1-4QMysteries', in F. García Martínez, ed., *Wisdom and Apocalypticism in the Dead Sea Scrolls and in Biblical Tradition* (Leuven: Peeters, 2003), pp. 69-88.

VanderKam, J.C., *Calendars in the Dead Sea Scrolls, Measuring Time* (London, Routledge, 1998).

VanderKam, J.C., 'Sources for the Astronomy in 1 Enoch 72-82', in C. Cohen, et al., eds., *Birkat Shalom: Studies in the Bible, Ancient Near Eastern Literature and*

Postbiblical Judaism Presented to Shalom M. Paul on the Occasion of His Seventieth Birthday (Winona Lake, Indiana: Eisenbrauns, 2008), pp. 965-78.

VanderKam, J.C., 'Calendars: Ancient Israelite and Early Jewish', in D.N. Freedman, ed., *Anchor Bible Dictionary, Volume 1* (New York: Doubleday, 1992), pp. 814-20.

Vitruvius, *Ten Books on Architecture*, trans. I.D. Rowland (Cambridge: Cambridge University Press, 1999).

Wacholder, B.Z. and D.B. Weisburg, 'Visibility of the New Moon in Cuneiform and Rabbinic Sources', *Hebrew Union College Annual*, (1971), Vol. 42, pp. 227-42.

Weinstock, S., ed., *Catalogus Codicum Astrologorum Graecorum, Volume 9:1* (Brussels: Lamertin, 1951).

Williams, C., 'Signs from the Sky, Signs from the Earth: The Diviner's Manual Revisited', in J.M. Steele and A. Imhausen, eds., *Under One Sky: Astronomy and Mathematics in the Ancient Near East* (Münster: UgaritVerlag, 2002), pp. 473-85.

Wise, M.O., 'Thunder in Gemini: An Aramaic Brontologion (4Q318) From Qumran', in *Thunder in Gemini And Other Essays on the History, Language and Literature of Second Temple Palestine, Journal for the Study of the Pseudepigrapha Supplement Series 15* (Sheffield: Sheffield Academic Press, 1994), pp. 13-50.

Wise, Michael, Martin Abegg and Edward Cook, eds., *The Dead Sea Scrolls: A New Translation* (New York: HarperCollins, 2005).

Wise, Michael, 'A Horoscope Written in Code (4Q186)', in Michael Wise, Martin Abegg and Edward Cook, eds., *The Dead Sea Scrolls: A New Translation* (New York: HarperCollins, 2005), pp. 275-77.

Vermes, Geza, *The Complete Dead Sea Scrolls in English* (London: Penguin, 1997), pp. 26-90.

Reshaping Karma: an Indic Metaphysical Paradigm in Traditional and Modern Astrology

Martin Gansten

Despite its Hellenistic origins, horoscopic astrology has nowhere gained as firm and lasting acceptance as in India. It is argued that this acceptance is conditioned by the conformity of astrological practice — comprising both descriptive and prescriptive aspects — to the doctrine of action or karman *central to the Indic religions. The relation of astrology to this doctrine is examined with regard to questions on causality, determinism and moral freedom. Traditional conceptions of* karman *are then contrasted with fin-de-siècle Theosophical notions of 'karma' as a fundamentally evolutive, spiritual force, used to redirect the practice of astrology from prediction towards esoteric interpretation. It is noted that this modern development constitutes a reversal of the European medieval and Renaissance compromise between theology and astrology.*

Horoscopic astrology is now commonly agreed to have been invented or discovered — depending on what view one takes of its legitimacy — in Hellenistic Egypt, around the second century BCE.[1] Some three centuries later it had made its way into northern India.[2] The Indian linguistic evidence clearly demonstrates the foreign origins of the discipline: from the earliest times, Sanskrit astrological literature abounds with Greek technical terms, in much the same way that modern books on computers and computer science in almost any language abound in (American) English jargon. The very word for astrology itself — *horā* — is of Greek derivation (ὥρα), as are the terms for its core technical concepts: *kendra* (κέντρον, angle), *paṇaphara* (ἐπαναφορά, succedent), *āpoklima* (ἀπόκλιμα, cadent), *trikoṇa* (τρίγωνον, trine), *meṣūraṇa* (μεσουράνημα, midheaven), *kemadruma* (κενοδρομία, being void of course), and so on. Most of these are terms for

1 David Pingree, *From Astral Omens to Astrology: From Babylon to Bikāner* (Roma: Istituto Italiano per l'Africa e l'Oriente, 1997), [hereafter Pingree, *From Astral Omens to Astrology*], p. 21. I use the term 'horoscopic' here in the literal sense of containing or being based on the ὡροσκόπος (rising sign or degree), not in the wider sense of being 'based on the date of birth', as in Nicholas Campion, *The Dawn of Astrology* (London: Continuum, 2008), [hereafter Campion, *The Dawn of Astrology*], p. 75.
2 Pingree, *From Astral Omens to Astrology*, p. 33.

which no indigenous equivalents were ever coined, and which are used by Indian astrologers to the present day. Given these extra-Indian roots of horoscopic astrology, how do we explain the near-universal acceptance which it has enjoyed in Hinduism and in Indian culture generally for the best part of two thousand years?

From the time of its inception, astrology has served a twofold purpose. On one hand, it attempts to analyse and interpret the qualities inherent in a given point of space-time – for instance, the time and place of a person's birth. These qualities will manifest in whatever is begun or produced at this point, such as the unfolding of a human life. This is the basis of what we may call the *descriptive* role of astrology, which obviously presupposes a certain element of determinism or predictability. On the other hand, astrologers also advise on how to make best use of the qualities of space-time by undertaking or refraining from particular actions. In this *prescriptive* role, astrology clearly presupposes a certain measure of freedom to act on the part of the individual.

Early astrological texts are not much concerned with presenting detailed philosophical analyses of how or why astrology is supposed to work, focusing instead on practical rules and instructions. This has left later generations of astrologers great freedom to adapt their theoretical understanding of the art – what some might like to call its 'ideological superstructure' – while maintaining a high degree of continuity in terms of practice. In Europe and the Middle East, astrology has survived and, at times, flourished within the physical and intellectual boundaries of the monotheistic and absolutist Abrahamic religions, all rather different from the pluralistic and polytheistic milieu of its origin. But nowhere has astrology blended so seamlessly with the dominating *Weltanschauung*, with the metaphysical assumptions and ritual practices, as in India. The explanation, I believe, lies in the ubiquitous Indian doctrine of *karman* or action. (I use the full Sanskrit stem form *karman* here, rather than the usual 'karma', to distinguish the concept found in the Indic religions from its modern reinterpretations.)

Like astrology, *karman* – a fundamental tenet of Hinduism, Buddhism and Jainism – takes the middle ground between the doctrines of fate (*daiva*) and free will (*puruṣakāra*). Although teachings on *karman* within these traditions vary in detail, common to all is the belief that the moral value of actions performed in previous lifetimes determines the individual's present circumstances, thereby creating the framework within which new action is performed, and so on *ad infinitum*. In itself, this process has neither beginning nor end. *Karman* is inseparable from the cycle of *saṃsāra* or transmigration, of which it is the driving force; and *saṃsāra* constitutes a closed system without entry point or natural progression towards any final destination.[3]

3 A *locus classicus* for the beginninglessness of *karman* and *saṃsāra* in brahminical

In allowing for the interplay of fate and free will, *karman* thus offers an excellent theoretical model for astrology, more consistent with how the art as actually practised than an absolute fatalism or doctrine of divine predestination. Not surprisingly, this model has in fact been invoked by Indian astrological authors for more than 1,700 years. Two images are prominently used to describe the role of astrology in relation to *karman*: that of divinity writing a person's destiny on his forehead, and that of a lamp illuminating objects in a dark room. The two are often, but not always, combined. A few examples will suffice:

> The edict carrying the impact of previous action (*karman*), which was inscribed on one's forehead by the Creator, is revealed by this science even as a lamp [reveals] objects in utter darkness.[4] (c. 300)

> The ripening of good and evil action (*karman*) accumulated in another birth is revealed by this science, as a lamp [reveals] objects in darkness.[5] (c. 550)

> That string of letters which was written by the Creator on one's forehead may be clearly read by an astrologer with the flawless eye of astrology (*horā*).[6] (c. 800)

> The row of letters which was written by the Creator on the tablet of men's foreheads in the world is truly revealed by an astrologer, and none other, with the flawless eye of astrology (*horā*). The ripening of what good or evil action (*karman*) was acquired in another birth is revealed by

tradition is Brahmasūtra 2.1.35 (*na karmāvibhāgād iti cen nānāditvāt*), refuting the objection that *karman* theory involves circular reasoning, the varying circumstances of life being conditioned by actions and actions again by circumstances of life. Śaṅkara comments: 'This is no fault [of reasoning], as transmigration (*saṃsāra*) is beginningless. It would be a fault if this transmigration had a beginning; but transmigration being beginningless, the existence of action (*karman*) and diversity of creation in the form of [both] cause and effect, like seed and sprout, is not self-contradictory' (*naiṣa doṣo 'nāditvāt saṃsārasya / bhaved eṣa doṣo yady ādimān ayaṃ saṃsāraḥ syād anādau tu saṃsāre bījāṅkuravad dhetuhetumadbhāvena karmaṇaḥ sargavaiṣamyasya ca pravṛttir na virudhyate*). Translations of this and other quotations in this paper are mine unless otherwise stated.

4 Vṛddhayavanajātaka 1.3:
yā pūrvakarmaprabhavasya dhātrī dhātrā lalāṭe likhitā praśastiḥ /
tāṃ śāstram etat prakaṭaṃ vidhatte dīpo yathā vastu ghane 'ndhakāre //

5 Laghujātaka 1.3:
yad upacitam anyajanmani śubhāśubhaṃ tasya karmaṇaḥ paktim /
vyañjayati śāstram etat tamasi dravyāṇi dīpa iva //

6 Sārāvalī 2.1:
vidhātrā likhitā yāsau lalāṭe 'kṣaramālikā /
daivajñas tāṃ paṭhed vyaktaṃ horānirmalacakṣuṣā //

this science through the order of planetary periods (*daśā*), as a lamp [reveals] pots and other kinds [of objects] in darkness.[7] (c. 1100-1400)

As is clear from these and similar passages, Sanskrit authors typically take a non-causal view of astrology: the stars 'reveal' (*vyañj-*) good and evil events to come, but do not cause them. The real cause is *karman*, action performed in previous lives, which 'ripens' (*pac-*) into events in the current lifetime. Future events themselves are compared to physical objects in a dark house: although not apparent to the unaided eye, they do exist and may be discovered by the proper method.

The image of such destined events as a divine edict written on man's forehead should be understood within its brahmanical context. There is no divine despotism involved in these decrees, which rather 'carry the impact of previous action': the Creator or Ordainer (*dhātṛ*, *vidhātṛ*) postulated by most forms of Hinduism is not Job's inscrutable autocrat, but the supervisor and guarantor of the workings of *karman*, and it is our own actions which mould our destinies. In fact, the concepts of a supreme God and of *karman* serve to justify each other: for while the moral law of *karman* may be more easily understood and accepted as the just judgments of an omniscient divinity than as a purely impersonal force, it also saves this divinity from the objection of partiality or caprice. The medieval theologian Śaṅkara compares the role of God to that of the rain: the rain makes crops grow, but some seeds will grow into barley and others into rice. Similarly, God allows the actions of each living being to ripen according to their varying qualities.[8]

Wilhelm Halbfass has argued that 'it can hardly be doubted' that this view of *karman* as the underlying mechanism of astrology was sometimes challenged, and that earlier times especially — when Indian astrology was significantly influenced by its Greek and Babylonian source traditions — saw 'a belief in an independent power of the stars to determine destiny, a power which can in no way be reduced to karma'.[9] This may seem a very reasonable supposition; but Halbfass produces no evidence from astrological texts to support it, and, to the best of my knowledge, there is

7 Horāmakaranda, introduction, verses 8-9 (reading *vyañjayatīha* for *vāṃcayatīha*):
varṇāvalī yā likhitā vidhātrā lalāṭapaṭṭe bhuvi mānavānām /
horādṛśā nirmalayā yathāvat tāṃ daivavid vyañjayatīha nānyaḥ //
yad anyajanmany aśubhaṃ śubhaṃ vā karmārjitaṃ tasya vipaktim etat /
vyanakti śāstraṃ hi daśākrameṇa ghaṭādijātaṃ tamasīva dīpaḥ //

8 Brahmasūtrabhāṣya ad 2.1.34.

9 Wilhelm Halbfass, *Karma und Wiedergeburt im indischen Denken* (Kreuzlingen: Diederichs, 2000), [hereafter Halbfass, *Karma und Wiedergeburt*], p. 239: 'Es kann aber kaum bezweifelt werden, daß dies nicht immer und nicht bei allen Lehrern oder Praktikern der Astrologie so gewesen ist. Es gab, insbesondere in älterer Zeit, auch den Glauben an eine selbständige, schicksalsbestimmende Macht der Gestirne, eine Macht, die keineswegs auf das Karma reduziert werden kann'.

none to be had. Even the rare astrological passages which seem to hint at some causal power inherent in the *grahas* ('planets' in the original sense of πλάνητες ἀστέρες, 'wandering stars' including the sun and moon) typically do so within the framework of *karman* theory. This is the case when, for instance, the Bṛhatpārāśarahorāśāstra, an astrological work of c. 600-800, states that God — here identified as Janārdana, or Viṣṇu — assumes the form of the planets to bestow on living beings the results of their actions (*karman*).[10] The *karman* model is often explicitly invoked, and never explicitly rejected, so that Halbfass' grudging recognition that 'astrology is to a certain degree reconciled with the doctrine of karma' seems so understated as to be quite misleading.[11]

The one passage which Halbfass does cite in support of his belief in an early doctrine of independent astrological causality is taken not from an astrological text, but rather from a *dharmaśāstra* or socio-religious code of law dating from the early centuries CE:

> When any [planet] is ill-placed for anyone, he should endeavour to worship that [planet]; [for] a boon was given to them by Brahmā: '[Having been] worshipped, you will worship [in return]'. The rise and fall of kings and the existence and annihilation of the world depend on the planets: therefore the planets are most worthy of worship.[12]

Taken in isolation, these verses could perhaps sustain the interpretation placed on them by Halbfass, although the evidence would have to be called circumstantial. They occur, however, alongside directions for the worship of other deities, in a text which explicitly upholds the doctrine of *karman*.[13] The same verses are also repeated with minor variations in the *Bṛhatpārāśarahorāśāstra*, which, as we have just seen, likewise supports the *karman* theory.[14] We must conclude, then, that if ancient India did indeed

10 Bṛhatpārāśarahorāśāstra 2.3: *jīvānāṃ karmaphalado graharūpī janārdanaḥ.*

11 Halbfass, *Karma und Wiedergeburt*, p. 240: 'Die Astrologie ist in gewissem Maße mit der Karmalehre versöhnt'.

12 Yājñavalkyasmṛti 1.307-308:
yaś ca yasya yadā duḥsthaḥ sa taṃ yatnena pūjayet /
brahmaṇaiṣāṃ varo dattaḥ pūjitāḥ pūjayiṣyatha //
grahādhīnā narendrāṇām ucchrāyāḥ patanāni ca /
bhāvābhāvau ca jagatas tasmāt pūjyatamā grahāḥ //

13 So, for instance, Yājñavalkyasmṛti 1.349:
daive puruṣakāre ca karmasiddhir vyavasthitā /
tatra daivam abhivyaktaṃ pauruṣam paurvadehikam //
'The accomplishment of an act (*karman*) depends on both fate and human effort. Of these, fate is the manifestation of effort [performed] with a previous body'.

14 Bṛhatpārāśarahorāśāstra 84.26-27:
yasya yaś ca yadā duḥsthaḥ sa taṃ yatnena pūjayet /
eṣāṃ dhātrā varo dattaḥ pūjitāḥ pūjayiṣyatha //
mānavānāṃ grahādhīnā ucchrāyāḥ patanāni ca /

know a belief in the 'power of the stars to determine destiny' independent of and rivalling the belief in *karman*, the textual evidence of that belief is yet to be produced. It may also be noted that astrology is only one of several divinatory arts historically practised in Indian culture, and that no causal link is normally assumed between the signs observed and the events foretold; rather, *karman* is the cause. A sixth-century text states: 'To travelling men, auspices (*śakuna*) proclaim the ripening of good and evil actions (*karman*) performed in other births'.[15]

Nevertheless, the verses just discussed do raise the question of whether religious acts such as the worship of deities — planetary or otherwise — can alter the future determined by *karman* and revealed by means of astrology. In theories on *karman*, actions performed in previous lifetimes are generally designated as 'accumulated' (*saṃcita*). Accumulated action is further divided into that portion which has begun to take effect (*prārabdha*) and that which has not (*aprārabdha*). The former is the *karman* determining the experiences of the present lifetime; the latter is stored up for lifetimes to come. Actions due to take effect in subsequent lifetimes may be neutralised by soteriological means; but the effects of *prārabdha* cannot be absolutely reversed, and it may therefore be seen as the more 'fated' aspect of *karman*. [16] And yet, the most basic assumption of *karman* theory is that we do have the freedom to choose between various courses of action, and are morally responsible for these choices. So which aspects of a human life are determined by actions in previous lifetimes, and can those aspects be at all influenced by actions in the present?

---

bhāvābhāvau ca jagatāṃ tasmāt pūjyatamā grahāḥ //
'When any [planet] is ill-placed for anyone, he should endeavour to worship that [planet]; [for] a boon was given to them by the Creator: '[Having been] worshipped, you will worship [in return]'. The rise and fall of men and the existence and annihilation of the worlds depend on the planets: therefore the planets are most worthy of worship'. The verses are not present in all editions of the text.

15 Yogayātrā 23.1:
anyajanmāntarakṛtaṃ puṃsāṃ karma śubhāśubham /
yat tasya śakunaḥ pākaṃ nivedayati gacchatām //

16 In Śaṅkara's words, 'as an arrow, already released from the bow to hit a mark, even after hitting it ceases its flight only with the exhaustion of the force generated, so the action (*karman*) generating the body, although directed towards the purpose of maintaining the body, continues [to produce effects] as before until the force of the impressions (*saṃskāra*) [caused by *karman*] is exhausted. But that same arrow unreleased, the force causing its movement not yet generated, though set to the bow is withdrawn; thus, actions which have not begun to take effect [but] remain in their own resting-place are rendered impotent by knowledge' (Bhagavadgītābhāṣya ad 13.23: *yathā pūrvaṃ lakṣavedhāya mukta iṣur dhanuṣo lakṣyavedhottarakālam apy ārabdhavegakṣayāt patanenaiva nivartata evaṃ śarīrārambhakaṃ karma śarīrasthitiprayojane nivṛtte 'py ā saṃskāravegakṣayāt pūrvavad vartata eva / sa eveṣuḥ pravṛttinimittānārabdhavegas tv amukto dhanuṣi prayukto 'py upasaṃhriyate tathānārabdhaphalāni karmāṇi svāśrayasthāny eva jñānena nirbījīkriyante*).

Let us first be clear that these are not questions confined to astrology. Hindu codes of law frequently prescribe acts of atonement or *prāyaścitta*, meant to avoid or mitigate any future retribution — in this lifetime or the next — for sins committed either knowingly or unknowingly.[17] Apparently, then, a conscious act of piety in the present is considered capable of counteracting or mitigating the results of past misdeeds, at least to some extent. When an evil event is foreseen, by astrology or any other means, and related by *karman* theory to some unknown sin committed in a previous existence, such an attempt to avert it is known as *śānti*, 'pacification' or 'propitiation'. The practice of *śānti* underscores the dual nature of both *karman* and astrology: on one hand, the accumulated actions of previous lives ripening into a destiny described by the horoscope; on the other, the chance of acting on one's knowledge of the stars to improve on one's natal prospects.

Seeking a balance between these two aspects, astrological authors have divided *prārabdha-karman* further. In the seventeenth century we find the encyclopedist Balabhadra discussing the arguments for and against what we may call the 'strong' view of karmic causality.[18] He quotes a previous author as saying: 'Not even the counsellor of the king of gods, who has direct knowledge of destiny, is able to alter the fate which someone is to experience'.[19] To this and similar statements, Balabhadra replies by making a distinction between 'firmly rooted' (*dṛḍha-mūla*) and 'loosely rooted' (*śithila-mūla*) *karman*, only the former of which gives rise to irrevocable 'fate'. A possible future event arising from less fixed *karman* can be counteracted, and herein lies the practical value of astrology.

Broadly speaking, however, it may be said that previous *karman* is believed to determine our experiences in the present lifetime — painful or pleasurable — along with our birth and, some would say, span of life.[20]

---

17 See, for instance, Manusmṛti 11.45-46:
akāmataḥ kṛte pāpe prāyaścittaṃ vidur budhāḥ /
kāmakārakṛte 'py āhur eke śrutinidarśanāt //
akāmataḥ kṛtaṃ pāpaṃ vedābhyāsena śudhyati /
kāmatas tu kṛtaṃ mohāt prāyaścittaiḥ pṛthagvidhaiḥ //
'The wise prescribe atonement for sin committed unintentionally, and some, because of indications in Revelation (*śruti*), even for [sin] committed wilfully. Sin committed unintentionally is washed away by study of the Veda; [sin] committed wilfully out of delusion, by atonements of various kinds'.

18 Horāratna, introductory chapter.

19 *yena tu yat prāptavyaṃ tasya vidhānaṃ sureśasacivo 'pi /*
yaḥ sākṣān niyatijñaḥ so 'pi na śakto 'nyathā kartum //

20 See, for instance, Yogasūtra 2.13: 'When the cause [i.e., *karman*] is present, its effects are birth, length of life, and experiences' (*sati mūle tadvipāko jātyāyurbhogāḥ*). Variations on this definition naturally exist, but most tend to include *āyus* or life span. Medical works such as the Carakasaṃhitā (3.3.28-38), however, argue that longevity depends on a combination of past and present action, and present a number of common-sense objections to the notion of an absolutely fixed span of

The *outcome* of our actions is therefore largely determined beforehand, but not the actions as such. And yet this is only half true: for our actions shape not only our external fortunes, but also our character; and our character, very often, determines our further actions. The earliest formulations of *karman* theory are aware of this. The Bṛhadāraṇyaka-Upaniṣad (c. eighth century BCE?) states:

> As one acts, as one lives, so he becomes. One who does good becomes good; one who does evil becomes evil. One becomes virtuous by virtuous action, evil by evil. Therefore they say, 'A man here is made of desire'. As his desire is, so will his intentions be; as his intentions are, so will he act; as he acts, so will he become.[21]

For this reason, *karman* may cause negative as well as positive 'spirals' — a notion found, for instance, in the Bhagavadgītā, which speaks of a reborn *yogin* being 'helplessly carried away' by the force of impressions from his previous lives until he 'treads the highest path', but also of God hurling evil men ever further down into rebirths in 'demonic wombs' (usually interpreted as lower species) until they 'tread the vilest path'.[22] These and similar passages have in fact led certain schools of Hindu thought to accept a doctrine of eternal transmigration, without possibility of liberation, for some souls.

Karman, then, is considered to influence both man's external fortunes and his internal qualities, or what we might loosely call his 'soul' (although, in the Hindu and Jaina view, our innermost being lies beyond the mutable character). This interest in the inner as well as the outer man is shared by both Greek and Indian astrology, which treat of a subject's mental

---

life. Rather, they advocate the idea of a maximum duration which may be cut short by overexertion, overeating or not eating enough, excessive copulation, illness wrongly treated, etc. Similar distinctions between maximum longevity (*paramāyus*) and 'untimely' or 'accidental' death (*akālamṛtyu, apamṛtyu*) are found in astrological works.

21 Bṛhadāraṇyaka-Upaniṣad 4.4.5: *yathākārī yathācārī tathā bhavati / sādhukārī sādhur bhavati / pāpakārī pāpo bhavati / puṇyaḥ puṇyena karmaṇā pāpaḥ pāpena / atho khalv āhuḥ kāmamaya evāyaṃ puruṣa iti / sa yathākāmo bhavati tatkratur bhavati / yatkratur bhavati tat karma kurute / yat karma kurute tad abhisaṃpadyate //*

22 Bhagavadgītā 6.43-45, 16.19-20:
tatra taṃ buddhisaṃyogaṃ labhate paurvadehikam /
yatate ca tato bhūyaḥ saṃsiddhau kurunandana //
pūrvābhyāsena tenaiva hriyate hy avaśo 'pi saḥ /
jijñāsur api yogasya śabdabrahmātivartate //
prayatnād yatamānas tu yogī saṃśuddhakilbiṣaḥ /
anekajanmasaṃsiddhas tato yāti parāṃ gatim //
tān ahaṃ dviṣataḥ krūrān saṃsāreṣu narādhamān /
kṣipāmy ajasram aśubhān āsurīṣv eva yoniṣu //
āsurīṃ yonim āpannā mūḍhā janmani janmani /
mām aprāpyaiva kaunteya tato yānty adhamāṃ gatim //

proclivities as matter-of-factly as they deal with his prospects in matters of health, prosperity, worldly power or love. But in the encounter with the Abrahamic faiths, and most particularly Christianity, the 'qualities of the soul' were to prove a stumbling-block.

Many authorities of the early Church, of course, rejected astrology in its entirety as incompatible with exclusive reliance on and reverence for God — a position reaffirmed by the Catholic catechism of recent years.[23] Justin of Caesarea (c. 100-c. 165) and Tertullian (c. 160-c. 220) both considered the art of astrology to have been discovered by fallen angels, and therefore to be condemned by God; and Augustine (354-430), himself a former student of astrology, similarly claimed that the predictions of astrologers come true because they are dictated by evil spirits.[24] Successive Church councils from the fourth century onwards anathematised practice of and belief in astrology; and the first Christian emperor (Constantine I, c. 272-337) threatened astrologers with death, while his son and successor (Constantine II, 316-340) vowed to have them ripped apart with iron claws.[25]

The late Middle Ages, however, saw a wider tolerance of astrology in Christian Europe, into which it had been re-imported along with other aspects of Greek science preserved and developed in the Islamic world. Although never universally accepted by religious authority in either culture, astrology played an important part in medieval physics and medicine. From the realm of theology, however, the stars were absolutely banned; and it was to this realm that the human soul belonged. 'The soul', preached Bernardino of Siena (1380-1444), 'is above the realm of the Moon, of Mercury, of Venus, of the Sun, of Mars, of Jupiter, of Saturn and of all the signs which are in them: it is above the 72 constellations'.[26] And a century and a half before, another Franciscan preacher, Berthold of Regensburg, had written:

> [The stars] have power over trees and over vines, over leaves and grasses, over vegetables and herbs, over corn and all such things; over the birds in the air, over the animals in the forests, and over the fishes in the waters and over the worms in the earth; over all such things that are under heaven, over them our Lord gave power to the stars, except over one thing. ... It is man's free will: over that no man has any authority except thyself.[27]

23 *Catechism of the Catholic Church* (London: Chapman, 1994), articles 2115-2116.
24 Campion, *The Dawn of Astrology*, p. 267 ff.
25 Benson Bobrick, *The Fated Sky: Astrology in History* (New York: Simon & Schuster, 2005), p. 83.
26 Quoted in Eugenio Garin, *Astrology in the Renaissance: The Zodiac of Life* (London: Routledge & Kegan Paul, 1983), p. 32. The somewhat opaque phrase 'of all the signs which are in them' may be a mistranslation for 'of all the [zodiacal] signs in which they are', but I have not seen the original text.
27 Quoted in Tester, *A History of Western Astrology* (Woodbridge: Boydell, 1987), p. 178.

The stars could be admitted to 'rule' the natural, sublunar world; but man's soul must be free to accept or reject divine grace and salvation, and must therefore be immune to astrology. This stance became a common if somewhat uneasy compromise between ecclesiastical and astrological teachings throughout the Renaissance. As we shall see, it is the exact opposite of the view held by the astrological reformers of modern times.

With the introduction of new scientific paradigms, interest in astrology declined drastically on the European continent during the seventeenth century. At the same time, the art was enjoying an unprecedented popularity in England; but a few decades into the next century, fashions had changed even here, and only the occasional enthusiast was left. It was not until the late 1880s that the first stirrings of a movement to popularise astrology were felt, a movement which was largely the creation of one man: William Frederick Allen, soon to be better known as Alan Leo (1860-1917). Leo's efforts proved successful in the way so common to popularising ventures: by altering the thing popularised to the point where one has to ask whether it is, in any meaningful sense, the same thing at all, or rather a new product marketed under an old label.

Astrology was only one of Leo's two great enthusiasms, the other being Theosophy as taught by Helena Blavatsky and, later, Annie Besant — teachings which in themselves were intended as a popularisation of the esoteric or 'occult' truths supposedly contained in all ancient religious traditions, although couched mainly in eastern terminology. Leo's life project was to unite the two by reinterpreting astrology as a spiritual doctrine, or, in the words of Wilhelm Knappich, to strip it of its scholastic-Aristotelic dress and shroud it in 'the shimmering magic cloak of Indian Theosophy' instead.[28] (The amount of Indian ideas actually contained in the Theosophical mélange is a point which we shall examine shortly.) 'There are two aspects of this Science', Leo wrote: 'the *exoteric* and the *esoteric*'.

> That side of Astrology which we call exoteric may be styled fatalism, fortune-telling, charlatanry — what you will: but the esoteric Astrology is that which reveals the soul of the Science, its divine aspect ... those whose minds are intuitive enough to catch the hidden significance of the esoteric side of Astrology know that it is part of THE MYSTERIES.[29]

---

28 'Denn die empirisch-praktischen Engländer, die schon längst der neuen Himmelsmechanik zugetan waren, zogen ihr einfach das scholastisch-aristotelische Kleid aus ... In dieser vereinfachten, aber etwas „nackten" Form wurde sie dann später von Alan Leo, H. S. Green, Sepharial u.a. in den schillernden Zaubermantel der indischen Theosophie eingehüllt ...', from Wilhelm Knappich, 'Placido de Titi's Leben und Lehre', *Zenit*, (1935), pp. 7-11.

29 Alan Leo, *Astrology for All* (New York: Cosimo Classics, 2006 [1910]), p. 293.

No highly developed powers of intuition are required to discern which of these two 'aspects' Leo valued more; and he did in fact admit quite openly that 'the esoteric side ... is the only part of the science that really interests me'.[30] There is no reason to doubt the sincerity of this statement; but neither can it be doubted that Leo's distaste for 'fortune-telling' was sharpened by the court cases brought against him — first in 1914 and again in 1917, only months before his death — on the very charge of telling fortunes. Following the first case, which was dismissed on technical grounds, Leo wrote:

> Let us part company with the fatalistic astrologer who prides himself on his predictions and who is ever seeking to convince the world that in the predictive side of Astrology alone shall we find its value. We need not argue the point as to its reality, but instead make a much-needed change in the meaning of the word and call Astrology the science of *tendencies*...[31]

The antipathy was mutual, and contemporary 'fatalistic' advocates of a mathematically rigorous, no-nonsense predictive astrology such as A. J. Pearce (1840-1923) had in fact already parted company with Leo and his fellow Theosophists, denouncing their metaphysical ideas as 'superstitious nonsense' and 'nauseating'.[32]

The same year, Leo and his wife formally merged astrology with Theosophy by founding the Astrological Lodge of the Theosophical Society. The previous year had seen the publication of Leo's most overtly Theosophical, and perhaps least popular, book: *Esoteric Astrology*, later described even by Charles E. O. Carter, president of the Astrological Lodge, as 'a big volume containing virtually nothing worth reading'.[33] It is useful, however, in giving us a clear idea of Leo's beliefs concerning the workings of karma and rebirth, 'the two pillars upon which all of Theosophical teaching rests'.[34] Here is Leo again:

> For one lifetime the soul will see everything from the point of view of Jupiter, and after death its experience will go to enrich the Individuality, making stronger within it the influence of Jupiter. Another Personality will follow it after an interval of rest in the heaven world, born under a different planet, intended to enrich another aspect of the Individuality; and when, after a succession of lives the time comes for the soul to be

30 Quoted in Patrick Curry, *A Confusion of Prophets* (London: Collins & Brown, 1992), [hereafter Curry, *A Confusion of Prophets*], p. 144.
31 Quoted in Curry, *A Confusion of Prophets*, p. 149.
32 Alfred John Pearce, 'Two Remarkable Horoscopes', *Star Lore*, (1897), p. 120.
33 Quoted in Curry, *A Confusion of Prophets*, p. 145.
34 Ronald Neufeldt, 'In Search of Utopia: Karma and Rebirth in the Theosophical Movement' in Ronald Neufeldt, ed., *Karma and Rebirth: Post Classical Developments* (New York: SUNY Press, 1986), [hereafter Neufeldt, 'In Search of Utopia'], p. 233.

> born again under Jupiter ... the remainder of the map [i.e., horoscope] of this Jupiter personality will be unlike that of the former one, because the soul will have grown, evolved, changed somewhat in the long interval, will have worked off some of its old karma, and will have made fresh, and the Divine Guardians of man will see that it is born in a different environment for the sake of obtaining increased experience. ...
>
> There is no other difference between souls than that which is due to the varied kind of experiences they have had in the past. The greatest sinner and the highest saint do not differ except in this, and in the fact that the saint is an old and experienced soul, whereas the sinner is relatively young and inexperienced as a soul. Birth in successive Personalities under new combinations of signs and planets, provides, astrologically speaking, the experiences required; and this will ensure that the sinner of to-day will be the great saint of the distant future.[35]

Analysing this passage, we find four interlinked themes. First of all, there is a *purpose* to our transmigratory existence; it is not mere blind mechanism or neutral fact. Second, this purpose is the gathering of *experience*, here described astrologically in terms of 'being born under' various planets and thereby experiencing the world from different points of view. Third, the accumulation of such variegated experience will bring about the *evolution* of the soul. Predicting the time and nature of the individual experiences is therefore of secondary interest at best, and at any rate cannot go beyond the identification of 'tendencies'; the important thing is how the experiences affect the soul. Fourth, this evolutive perspective presupposes *a beginning and an end* to the process of transmigration, with souls of varying age situated at different points in the spiritual curriculum.

This is all orthodox Theosophy, if such an expression is not an oxymoron. Ronald Neufeldt has rightly characterised Blavatsky's teachings on karma and rebirth as 'utopian', in the sense that progress or evolution constitutes their most important element: 'Indeed, rebirth becomes the means whereby progress is achieved under the sway of the law of karma'.[36] Such utopian belief in the inevitability of progress, in a chain of evolution where 'each fresh attempt is more successful than the previous one',[37] is highly characteristic of late nineteenth-century western thought, and reflects the profound impact of the ongoing industrial revolution. Blavatsky's metaphysical views are, in this respect, not dissimilar to the biological evolutionism of Darwin or the socio-political evolutionism of Spencer. They contrast sharply, however, with the ideas of *karman* and transmigration present in the Indic religions.

35 Alan Leo, *Esoteric Astrology* (Rochester: Inner Traditions, 1989 [1913]), [hereafter Leo, *Esoteric Astrology*], p. 104 f.
36 Neufeldt, 'In Search of Utopia', p. 247.
37 Quoted in Neufeldt, 'In Search of Utopia', p. 248.

I have already said that, in Indian thought, *saṃsāra* or the cycle of transmigration is a closed, beginningless system fuelled by *karman*. In Buddhism and Jainism this transmigratory existence is seen as a fundamental fact in itself, in no need of any further, underlying cause. In Hinduism, the world is typically considered as created by or emanating from God or the Absolute (*brahman*); but brahmanical theologians are also clear that the world is eternal, and its 'emanation' an ontological rather than a temporal relationship. The notion that creation serves some purpose is explicitly rejected; it is divine 'play alone'.[38]

There is no automatic progress built into the system of transmigration, nor is it a one-way road: the individual self or soul has been wandering through *saṃsāra* forever and, if left to the mechanism of *karman*, will continue to do so forevermore, raising or lowering itself by its own actions rather than evolving according to some grand design. There is no beginning, no end, no purpose, no progress; and 'experience', so far from leading to spiritual fulfilment, is the stuff from which the soul builds the walls that imprison it.[39] In view of these very different perspectives, it may well be asked how much the Theosophical teachings on karma incorporated into Leo's 'modernised' astrology really owe to India.

---

38 See Brahmasūtra 2.1.32-33 (*na prayojanavattvāt, lokavat tu līlākaivalyam*). Śaṅkara comments: 'Just as, in the world, the activities of a king or of a royal minister whose [every] desire is fulfilled take the form of mere play in places of amusement, without any other purpose in view ... so too the activity of Lord, disregarding any other purpose, by his very nature will take the form of mere play; for no other purpose of the Lord can be discerned either by reason or from Revelation (*śruti*) ...' (*yathā loke kasyacid āptaiṣaṇasya rājño rājāmātyasya vā vyatiriktaṃ kiñcit prayojanam anabhisandhāya kevalaṃ līlārūpāḥ pravṛttayaḥ*
krīḍāvihāreṣu bhavanti ... evam īśvarasyāpy anapekṣya prayojanāntaraṃ svabhāvād eva kevalaṃ līlārūpā pravṛttir bhaviṣyati na hīśvarasya prayojanāntaraṃ nirūpyamāṇaṃ nyāyataḥ śrutito vā saṃbhavati ...).

39 Although liberation (*mokṣa*) from the cycle of transmigration is considered possible only from certain positions within the world of *karman* and after lifetimes of striving (cf. the discussion of 'spiralling' *karman* above), such liberation is not achieved by action but, on the contrary, by the extinction of the effects of action (cf. note 16). Even the most 'spiritual' mode of experience within this world (*sattva-guṇa*) is considered potentially enslaving; cf. Bhagavadgītā 14.5-6:
sattvaṃ rajas tama iti guṇāḥ prakṛtisaṃbhavāḥ /
nibadhnanti mahābāho dehe dehinam avyayam //
tatra sattvaṃ nirmalatvāt prakāśakam anāmayam /
sukhasaṅgena badhnāti jñānasaṅgena cānagha //
'Purity (*sattva*), passion (*rajas*) and darkness (*tamas*) are the qualities born of [material] nature. They bind the imperishable embodied [self] to the body, O mighty-armed one. Among them, purity by its flawlessness is illuminating and free from suffering: it binds by the bonds of happiness and by the bonds of knowledge, O sinless one'.

As discussed earlier, the metaphysical conception of action so closely connected with astrology in India strikes a compromise between the ideas of absolute fate and absolute freedom. Leo and his followers similarly saw karma as an alternative to the 'fatalism' which they, rightly or wrongly, imputed to their more conservative — and, it must be said, often more technically astute — astrological colleagues. But while Indian astrology remained, and still remains today, a primarily predictive discipline, Leo, as we have seen, was highly disparaging of 'the predictive side of Astrology'.

Intriguingly, however, only statements about external events counted as 'prediction' with the Theosophical astrologers: delineating a person's character or mental qualities was seen as perfectly legitimate, indeed often as the *only* legitimate use of astrology, although it is difficult to see how this is different from prediction — particularly assuming that, in Leo's favourite catch-phrase, 'Character is Destiny'. The solution to the conundrum no doubt lies in the relation of astrology to the predominant ideology of the age. In pre-modern Europe, this meant Christianity as defined by the Church; and the principal domain of the Church was the human soul — a monopoly not to be infringed on. The soul therefore had to be safeguarded from planetary influences. In the late nineteenth and early twentieth centuries, by contrast, Science had largely replaced the Church as arbiter of truth; and the domain of Science, just as jealously guarded, was matter and mechanistic causality. The soul, if indeed it existed, was of little interest to Science, and astrologers were therefore free to expound upon it, as long as they did not stake any claim in the world of concrete and measurable results.

It was Dane Rudhyar (1895–1985), another highly influential Theosophist astrologer, who, two decades after Leo's demise, brought this 'tendency' to its logical conclusion by claiming that accurate prediction is not only impossible in practice, but actually undesirable:

> Besides, why should events be foretold accurately? The coefficient of inaccuracy is the coefficient of freedom. ... And to be free means always somewhat *not to know*; it is the coefficient of inaccuracy. It is based on the courage to go forth while not knowing the future.
>
> That is why spiritual teachers or 'Masters" — whatever they be — *never* compel, *never* show the exact future of any action undertaken. For to do so would be to rob a man of his creative freedom and his creative initiative. What man can do is so to understand the past, so to grasp the full significance of the seed-form of his being and destiny (birth-chart), that he is fully prepared to meet any future — to meet it significantly, with courage, with understanding and from such a "formed" view-point that all events are seen as beautiful. This is the creative and the radiant life of fulfillment.[40]

40 Dane Rudhyar, *The Astrology of Personality* (New York: Lucis Publishing, 1936), p. 460 f.

In short, where ignorance is creative fulfilment, 'tis folly to be wise. Astrology has done a volte-face: it no longer looks forwards in an attempt to predict and, if possible, control the future, but rather backwards, trying to find symbolic meaning in what has led up to the present. Naturally, karma — unless it is to be discarded altogether — has to be similarly re-defined as pertaining only to the past; and Rudhyar does precisely that: 'The inertia of the past (karma)', he says, 'makes the mind unable to clearly see the new possibilities for action and thought (dharma) which the birth-situation actually contains'.[41] Incidentally, we have here another casualty of astrological newspeak. *Dharma* is a versatile Sanskrit term meaning, in different contexts, law, duty, virtue (both in the sense of inherent nature and in the sense of morality) or religion; but it does *not* mean 'new possibilities for action and thought'. *Dharma* imposes restrictions on our choice of action rather than widening it.[42] Rudhyar, however, was in need of an antithesis to his concept of karma, and perhaps could not resist one that rhymes.

From this perspective of creative and courageous inaccuracy, karmic bondage lies in the past, while the future holds unlimited evolutional potential. The present, to Rudhyar, is far more than a fleeting moment: it is the dividing line between good and evil.

> Evil is essentially the refusal to move toward the future. It is to accept the repetitive inertia of past choices as inevitable or too powerful to oppose. It is to succumb to karma, instead of using what the past has produced as a floor against which to rebound, and of investing this rebounding with a creative, future-engendering meaning.[43]

We may not be entirely sure what a 'future-engendering meaning' is, but it is clear that the future has become, in C.S. Lewis' phrase, 'a promised land which favoured heroes attain — not ... something which everyone reaches at the rate of sixty minutes an hour, whatever he does, whoever he is'.[44] The role of astrology in relation to this view of time and karma is somewhat vague; but to 'grasp the full significance of the seed-form of one's being and destiny' in the form of the natal horoscope obviously does not entail any

41 Dane Rudhyar, 'Transmutation of Karma into Dharma', in Virginia Hanson, Rosemarie Stewart and Shirley Nicholson, ed., *Karma: Rhythmic Return to Harmony* (Wheaton: Quest Books, 1990 [1975]), [hereafter Rudhyar, 'Transmutation of Karma into Dharma'], p. 232.

42 The word is still used in something like its classical sense in Leo, *Esoteric Astrology*, p. 26: 'Saturn is the planet of "Dharma," duty or obligation, for every human creature'. In traditional Indian astrology, however, *dharma* is seldom mentioned except as a name for the 9th house (*bhāva*, *sthāna*) of the horoscope, and is not specifically related to Saturn.

43 Rudhyar, 'Transmutation of Karma into Dharma', p. 241 f.

44 Lewis, Clive Staples, *The Screwtape Letters*, p. 130.

foreknowledge of what actual events that seed will ripen into. The writing on the forehead remains obscure.

In conclusion, we see that the classical Indic concept of *karman* and the modern Theosophical notion of 'karma' have served to embed astrology within two rather dissimilar metaphysical structures. In India, an astrological praxis of Hellenistic origin, at once descriptive and prescriptive and spanning both internal and external aspects of human life, was successfully merged with a theory of action as the ultimate force shaping physical events and mental qualities in a beginningless cycle of rebirth with no other purpose or design. In the western world at the turn of the last century, where no such unifying theory existed, a hybrid version of karma, centred around the idea of constant progress or evolution so characteristic of the period, was used to reinforce the dichotomy between the subjective and objective spheres, and to steer students of astrology firmly away from the lower or exoteric astrology of 'fatalistic prediction' and 'fortune-telling' so hateful to the modern mind, directing them instead towards the higher, esoteric realm of spiritual symbolism.

Bibliography

Ancient and classical works

Bhagavadgītā, Bhagavadgītābhāṣya, *Śrīmadbhagavadgītā: Śrīmacchaṅkarabhāṣyeṇa Ānandagirikṛtavyākhyāyujā saṃvalitā*, ed. Wāsudev Laxmaṇ Shāstrī Paṇśīkar (Bombay: Nirṇaya-Sāgar Press, 1936).

Brahmasūtra, Brahmasūtrabhāṣya, *Brahmasūtraśāṅkarabhāṣyam*, eds. Rāmachandra Shāstrī Dhūpakar & Mahādeva Shāstrī Bākre (Bombay: Nirṇaya-Sāgar Press, 1904).

Bṛhadāraṇyaka-Upaniṣad, *The Early Upaniṣads*, text and trans. Patrick Olivelle (Oxford: Oxford University Press, 1998).

Bṛhatpārāśarahorāśāstra, *Brihat Parasara Hora Sastra*, Volume I-II, text and trans. Rangachari Santhanam & Gouri Shankar Kapoor (Delhi: Ranjan Publications, 1984-1991).

Carakasaṃhitā, *Caraka Saṃhitā*, Volume I-II, text and trans. Ram Karan Sharma & Bhagwan Dash (Varanasi: Chowkhamba Sanskrit Series, 1983-1988).

Horāmakaranda, *Hora Makarand*, text and trans. H. K. Thite (New Delhi: Sagar Publications, s.a.).

Horāratna, *Hora Ratnam*, text and trans. Rangachari Santhanam (New Delhi: R. Santhanam Associates, 1995).

Laghujātaka, *Śrīmadvarāhamihirācāryaviracitaṃ Laghujātakam*, ed. Laṣaṇalal Jha (Varanasi: Krishnadas Academy, 1983).

Manusmṛti, *The Manusmṛti: With the Commentary Manvarthamuktāvali of Kullūka*, ed. Nārāyaṇ Rām Āchārya (Bombay: Nirṇaya-Sāgar Press, 1946).

Sārāvalī, *Saravali*, text and trans. Rangachari Santhanam (New Delhi: Ranjan Publications, 1983).

Vṛddhayavanajātaka, *Mīnarājaviracitaṃ Vṛddhayavanajātakam*, ed. David Pingree (Baroda: Oriental Institute, 1976).

Yājñavalkyasmṛti, *Yājñavalkyasmṛti of Yogīśvara Yājñavalkya*, ed. Nārāyaṇ Rām Āchārya (Bombay: Nirṇaya-Sāgar Press, 1949).
Yogasūtra, *Pātañjalayogasūtrāṇi*, ed. Rajaram Shastri Bodas (Bombay: Government Central Press, 1917).
Yogayātrā, *Yogayātrā*, text with Sanskrit-Hindi commentary, Satyendra Miśra (Vārāṇasī: Krishnadas Academy, 1999).

Modern works

Bobrick, Benson, *The Fated Sky: Astrology in History* (New York: Simon & Schuster, 2005).
Campion, Nicholas, *The Dawn of Astrology* (London: Continuum, 2008).
Catechism of the Catholic Church (London: Chapman, 1994).
Curry, Patrick, *A Confusion of Prophets: Victorian and Edwardian Astrology* (London: Collins & Brown, 1992).
Garin, Eugenio, *Astrology in the Renaissance: The Zodiac of Life* (London: Routledge & Kegan Paul, 1983).
Halbfass, Wilhelm, *Karma und Wiedergeburt im indischen Denken* (Kreuzlingen: Diederichs, 2000).
Knappich, Wilhelm, 'Placido de Titi's Leben und Lehre', *Zenit*, (1935), pp. 7-11.
Leo, Alan, *Esoteric Astrology* (Rochester: Inner Traditions, 1989 [1913]).
Leo, Alan, *Astrology for All* (New York: Cosimo Classics, 2006 [1910]).
Lewis, Clive Staples, *The Screwtape Letters* (London: Fontana Books, 1971 [1942]).
Neufeldt, Ronald, 'In Search of Utopia: Karma and Rebirth in the Theosophical Movement', in Ronald Neufeldt, ed., *Karma and Rebirth: Post Classical Developments* (New York: SUNY Press, 1986), pp. 233-55.
Pearce, Alfred John, 'Two Remarkable Horoscopes', *Star Lore*, (Aug. 1897), p. 120.
Pingree, David, *From Astral Omens to Astrology: From Babylon to Bikāner* (Roma: Istituto Italiano per l'Africa e l'Oriente, 1997).
Rudhyar, Dane, *The Astrology of Personality: A Reformulation of Astrological Concepts and Ideals in Terms of Contemporary Psychology and Philosophy* (New York: Lucis Publishing, 1936).
Rudhyar, Dane, 'Transmutation of Karma into Dharma', in Virginia Hanson, Rosemarie Stewart and Shirley Nicholson, eds., *Karma: Rhythmic Return to Harmony* (Wheaton: Quest Books, 1990 [1975]), pp. 231-41.
Tester, Jim, *A History of Western Astrology* (Woodbridge: Boydell, 1987).

Néladóracht: Druidic Cloud-Divination in Medieval Irish Literature

Mark Williams

Irish literature in the high and late Middle Ages includes a large body of sagas set in the pre-Christian past, in which druids often feature as important characters. In a number of texts, druids perform a type of divination from the clouds, termed néladóracht. *This paper investigates the origins of the literary topos of druidic nephelomancy, arguing that it is ultimately dependent on astrology rather than any native, pre-Christian custom.*

The druids were the magico-religious intelligentsia of at least some of the Celtic-speaking peoples of north-western Europe in the centuries immediately either side of the birth of Christ, though they persisted rather longer in Ireland. As mysterious and evocative figures, they developed an unusually vivid and long-lasting cultural 'afterlife', extending from the early modern period down to the present, as Ronald Hutton has eloquently analyzed.[1] Their reputation as wise natural philosophers — one of a number of conflicting associations they possess in the classical texts which describe them — is belied by the lack of concrete details about the precise scope of their much-vaunted wisdom.

One of the recurring aspects of modern, popular envisionings of the philosophical wisdom of the druids is the ascription to them of some form of astrological lore. This is based very largely on a single statement by Julius Caesar in his *Gallic War*, describing the curriculum of the druids of Gaul in the first century BCE. According to Caesar, Gaulish druids 'hold long discussions about the heavenly bodies and their movements, the size of the universe and of the earth, the physical constitution of the world, and the power and properties of the gods; and they instruct the young men in these subjects'.[2] Indeed, so strong is the imaginative hold of this passage that one can now purchase imaginative volumes with titles like *The Lost Zodiac of the Druids*, *The Handbook of Celtic Astrology* and so on, none of which have the slightest historical validity. However, it is my contention in this article that

1 R. Hutton, *Blood and Mistletoe: the History of the Druids in Britain* (Yale University Press, New Haven & London, 2009).
2 Caesar, *De Bello Gallico*, vi, 14 (my translation).

this impulse to credit druids with some kind of divinatory, native sky-lore is not merely a modern phenomenon, but rather was shared by the composers of Irish vernacular saga-texts in the high Middle Ages, who occasionally depict druids and other pre-Christian prophets engaging in a form of cloud-divination, or *néladóracht* in the Irish language. The remainder of this article will explore this literary topos, and try to explain its emergence in medieval Ireland.

Ireland was the one place in Europe in which druids remained of interest and cultural importance during the Middle Ages, as common figures in the complex body of saga-narratives, set in the pre-Christian past, which were composed in Irish from the end of the seventh century onwards. Ireland was largely converted to Christianity in the late fifth and early sixth centuries CE, as a result of the mission associated with St Patrick, although we know that Christians, probably including slaves captured in Roman Britain, were already present in the island shortly after the year 400CE.[3] However, we can safely assume that the pre-Christian Irish did indeed have historical druids, like other Celtic speaking peoples. Native penitentials — Christian tracts giving the various penances one should do for every conceivable variety of sin — and law texts on social status make it clear that druids of a sort were still a going concern in Irish society until the turn of the eighth century, when they seem quietly to have disappeared.[4] Between the mid-fifth century and the beginning of the eighth, therefore, the historical druids seem to have endured a steep lowering of their social status, entering a kind of twilight decline in a rapidly Christianizing, and then Christian, Ireland.

However, at roughly the same time as real druids were finally fading permanently into history, an immensely vivid and inventive vernacular literary culture began to flower, which delighted in setting tales and sagas in an imagined version of the pre-Christian past of several hundred years previously. Druids play a prominent part in these tales as counselors to kings, magicians, judges and prophets, but the relationship between these literary druids and their historical forebears is complex and problematic. During the last thirty-to-forty years, scholars have had a revolution in the way that the cultural thought-world and artistic priorities of the literati of medieval Ireland are envisaged. There was once a tendency, especially in the late nineteenth century, to look at sagas from medieval Ireland as though they were tracts describing the culture of the pre-Christian Irish Iron Age in historically-reliable terms. Among scholars in the field, this tendency existed in more and less sophisticated versions, and is still to be

3 For the historical narrative of Ireland's conversion, see T. M. Charles-Edwards, *Early Christian Ireland* (Cambridge University Press, Cambridge, 2000), [hereafter Charles-Edwards, *Early Christian Ireland*], pp. 183ff.

4 See F. Kelly, *A Guide to Early Irish Law* (Dublin: Dublin Institute for Advanced Studies, 1988), pp. 60-61, and cp. *The Irish Penitentials,* ed. L. Bieler (Dublin: Dublin Institute for Advanced Studies, 1975), p. 160.

seen in many popular books about the ancient Celts, which tend to neglect current research in favour of the sepia wash of romance. But within the academy, we no longer think of these medieval sagas as accurate portrayals of life in pre-Christian Ireland. Rather, we recognise them as *constructed*: that is, as deliberate artistic creations — drawing on inherited material, to be sure, but adapting it freely to provoke, stimulate and entertain their medieval, Christian audience. (So standard has this consensus become in the field, it is sometimes termed the 'new orthodoxy'.) So we cannot look at the activities of druids in Irish medieval literary texts and take them naively as anthropologically accurate transcriptions.

We work today with an awareness that medieval Ireland was culturally highly creative and linked to the wider world of Christendom, not an isolated backwater brooding on a shadowy and legend-filled past. We also see vernacular saga literature emerging fundamentally from the monastery: the sagas, with their shapeshifting, pagan gods and ripe sexuality, are nevertheless products of a Christian intellectual elite who were familiar with Latin learning and steeped in the Bible and the biblical exegesis of the Church Fathers. Such learning affected the shaping of literary narratives at a profound level.[5]

In medieval Ireland, there were two main words for druid, one in each of the two languages of the literati. The first was Old Irish *druí*, of which the plural was (confusingly for modern English speakers) *druid*, and which derives from Common Celtic **dru-wids*, meaning 'thorough knower', or possibly 'oak-knower' — the first element is disputed. The second was Hiberno-Latin *magus*, meaning 'druid, wizard, magian, magician', and, as an important and obviously related meaning, 'one of the Magi of the Gospel according to St Matthew'. This latter meaning will be highly significant for my argument in this article.[6] These two terms are used essentially interchangeably, the choice being dependent on the main language of the text in question.

In the earlier sagas, which date from the seventh to the eleventh centuries, druids are depicted as divining the future and prophesying by several methods. In some texts they are capable of a kind of eerie clairvoyance, often termed *imbas for-osna*, or 'the embracing vision which enlightens'. In the eighth century saga, *Aided Chonchobuir*, for example, druids perceive the Crucifixion from the other end of Europe, seeing it in visionary 'real time', as it were; in the seventh century Latin *Vita Patricii*,

5 For an excellent discussion of the intellectual contexts of vernacular literary activity in early medieval Ireland, see D. Ó Cróinín, 'Ireland, 400-800' in D. Ó Cróinín, ed., *A New History of Ireland 1: Prehistoric and Early Ireland* (Oxford: Oxford University Press, 2005), pp. 182-88; for a more polemical position, see K. McCone, *Pagan Past and Christian Present in Early Irish Literature* (Maynooth: An Sagart, 2000), [hereafter McCone, *Pagan Past*].

6 In what follows, *magi* (lowercase) refers to wizards or druids, and *Magi* (capitalised) refers to the Wise Men of the second chapter of Matthew's Gospel.

ascribed to Muirchú, the druids (termed *magi*) prophesy the coming of the saint. Elsewhere, they are able to practice a form of prophetic psychometry: in the originally ninth century saga *Longes mac nUislenn*, the Ulster druid Cathbad is able to foresee the adult appearance and disastrous career of the tragic heroine Derdriu, simply by placing his hand on her mother's belly while Derdriu herself is still in the womb. Other sagas speak of druidic knowledge of lucky and unlucky days, and of dream-visions induced by wrapping oneself in the hide of a slaughtered bull before going to sleep. Any and all of these may genuinely reflect pre-Christian druidic practice, though we cannot be certain. None of these, however, are astrological.

Nevertheless, Celtic specialists have not been immune to the lure of imputing astrological know-how to ancient Irish druids. Fergus Kelly and James Carney — great scholars both — state straightforwardly that divination from celestial phenomena formed one of the roles of the Irish druid. Kelly tells us: 'The druid (Old Irish *druí*) was priest, prophet, *astrologer*, and teacher of the sons of nobles'.[7] Carney writes: 'In Latin writing, *druí* is translated *magus*, and his role is that of necromancer and watcher of the heavens'.[8] The confidence of both these statements should make us pause, though Carney's 'watcher of the heavens' perhaps shows a justified wariness about using the word 'astrologer'. On the one hand, those elements of the Irish depictions of druids which echo the Classical sources can be seen as confirming the latter, and thus being good evidence for the historical reality of a particular druidic custom. Kelly and Carney's perspectives implicitly follow this line of reasoning.

Amongst the Classical sources, there are indeed passages which support the idea that druids in continental Celtic societies studied the stars: we have already heard from Julius Caesar on Gaul. That passage, quoted above, may simply be historically accurate — Caesar had, after all, a pressing need to understand where power lay in Gaulish society. But, on the other hand, doubts have been cast on the accuracy of Caesar's ethnography of the region, as he may have been drawing on the lost testimony of the Syrian philosopher Posidonius here, and thus it is not certain that he was writing from personal experience when it came to specific details of this kind.[9] Indeed, the ascription of sophisticated natural philosophy to the druids may represent a projection of familiar Pythagoreanism onto a barbarian caste whose customs were largely unknown.

7 F. Kelly, *A Guide to Early Irish Law* (Dublin: Dublin Institute for Advanced Studies, 1988), pp. 59-60 (emphasis mine).

8 J. Carney, 'Language and literature to 1169', in D. Ó Cróinín ed., *A New History of Ireland, Volume I: Prehistoric and Early Ireland* (Oxford: Oxford University Press, 2005), p. 451 [hereafter Carney, 'Language and literature to 1169'].

9 The inherent historical accuracy of Posidonius' lost testimony is of course as difficult to assess as his probable influence on Caesar and other classical writers on the druids. See D. Nash, 'Reconstructing Posidonius's Celtic Ethnography', *Britannia,* (1976), Vol. 7, pp. 112-36.

Thus we must approach classical texts describing druidic star-knowledge with some wariness, despite their aura of plausibility. This applies particularly to their use as comparanda for references to druids in the Irish medieval literature, especially when one tries to create a picture of the practices of the historical druids thereby. In fact, there is only one reference from early medieval Ireland to a druid studying the stars, found in the mid-seventh century *Vita Prima S. Brigitae*, the earliest Latin *Life of St Brigit*, Ireland's mother-saint. Born to the slave-girl of a druid, Brigit's birth is accompanied by miraculous phenomena, which the druid observes in person:

> One night this druid (*magus*) was keeping watch, as was his custom, contemplating the stars of heaven, and throughout the entire night he saw a blazing column of fire rising out of the hut in which the slave-girl (*ancilla*) was sleeping with her daughter, and he called to him another man, and he saw the same thing.[10]

It is this very passage that Kelly cites as his evidence that Irish druids were 'astrologers'. Again, it is possible that this was indeed the case, and that these lines, written at a time when druids seem still to have existed in Ireland, do preserve genuine information about them. But if so, this is the *sole* passage to ascribe this kind of activity to an Irish druid, and a number of objections can be raised. First — in an excellent example of how interpretations of medieval Irish images of druids can be unconsciously coloured by our knowledge of the Classical accounts — it is not clear whether watching the stars through the night is to be taken as a custom of *magi* in general, or just of this particular *magus.* Secondly, if we do take the passage as implying that one of the roles of the druidic class was to study the stars, once again we cannot necessarily take this as reliable historical information. Finally, the detail is part of the narration of a natal *mirabilium*, and is, I suspect, intended to be taken as one of a series of deliberate echoes in the passage of Matthew's account of Christ's nativity. The pillar of fire over the hut clearly recalls the star of Bethlehem over the stable, and just as Matthew has the Magi following a star, the author of the *Vita Prima* has one *magus* scanning the stars.[11]

Matthew's Gospel does not make it explicit whether the Magi were astrologers, of course, but from early on Irish churchmen had one weighty source available to them which strongly stated that they were — namely the

10 *Vita Prima S. Brigitae*, §8, ed. J. Colgan, *Triadis Thaumaturgae ... Acta* (Louvain, 1647), p. 528 (my translation).

11 Catherine McKenna notes that the term *ancilla*, used of Brigit's mother, is the term Mary applies to herself in Luke 1:38; see C. McKenna, 'Between Two Worlds: Saint Brigit and Pre-Christian Religion in the *Vita Prima*', in J. F. Nagy, ed., *Identifying the 'Celtic': the Celtic Studies Association of North America Yearbook 2* (Dublin: Four Courts Press, 2002), [hereafter McKenna, 'Between Two Worlds'], p. 70.

Etymologiae, Isidore of Seville's great seventh-century encyclopedia, which we know was highly esteemed in Ireland within a few decades of its composition.[12] Isidore devotes a section of the *Etymologiae* to the subdivisions of *magi* as sinister and diabolically-inspired pagan diviners, drawing, as usual, on a wide variety of biblical and classical sources. Those who predict the future by the stars form an important category:

> *Astrologi* are so called because they make auguries from the stars. For they describe the births of human beings by means of the twelve signs of heaven, and through the movement of the stars they attempt to foretell the habits, roles, and fates of those who are born — that is, who will have been born in what sort of sign, or what fate he who is born may have in life. These are they who are commonly called *mathematici*. But originally these same star-interpreters were termed *magi*; thus it is said that they were the ones in the Gospel who announced that Christ had been born.[13]

It has long been recognised that the *magi* of Muirchú's *Vita Patricii*, clearly meant to be druids, were deliberately paralleled to the wicked priests and magicians of the Old Testament; and given Isidore's tremendous influence upon Irish learning from the mid-seventh century, it would be unsurprising to find Brigit's hagiographer drawing upon the former's description of pagan magicians to flesh out his *magus*. Thus, we are not obliged to take the *Vita Prima* at face value as evidence — the *only* evidence — for the astrological skills of Irish druids, as there is another possible and plausible explanation. Druidic astrology, at least in Ireland, is likely to be a mirage.

Carney described the historical druids as 'watchers of the sky', which is prudently imprecise on the issue of what it was that they were actually scrutinizing. But in the eleventh or twelfth century, a new skill — cloud divination — seems to have been added to the divinatory repertoire of the literary druid. Cloud-divination is both reminiscent of astrology and yet dissimilar to it; I aim to show below that its appearance as a druidic skill in literary texts in this period reflects developments in the meaning of the word *magus* in Ireland, and in particular its semantic superimposition of the senses 'pagan Irish diviner', on the one hand, and 'heaven-scanning, Christ-child-visiting wise man', on the other.

Examples of cloud-divination occur in at least five surviving vernacular texts of the high and late Middle Ages (from the twelfth century to the

12 See Carney, 'Language and literature to 1169', p. 390, where he states that the *Etymologiae* was in general use among Irish writers by the middle of the seventh century. As Carney notes, 'so esteemed were the 'Etymologies', especially, that the Irish referred to him affectionately as *Issidir in chulmin* ('Isidore of the summit', i.e., of the summit of learning)'.

13 Isidore of Seville, *Etymologiae*, in W. M Lindsay, ed., *Isidori Hispalensis Episcopi Etymologiarum Sive Originum Libri XX*, 2 Volumes (Oxford: Oxford University Press, 1911), [hereafter Isidore, *Etymologiae*], viii, 9, 22-5 (my translation).

fifteenth), of which I am going to discuss three here. These are a hagiographical work, the so-called 'Irish Life of Columba', plus two saga-narratives, *Acallam na Senórach* and the Stowe version of *Táin Bó Cúailnge*.[14] I will describe each of these texts briefly here in turn as they are unlikely to be familiar to non-specialists.

The 'Irish Life of Columba' is a Middle Irish *vita* of St Columba or Colum Cille of Iona, Ireland's third great saint after Patrick and Brigit, who died in 597CE. (It is not to be confused with Adomnán's much more famous *Vita Sancti Columbae*, produced on Iona around a century after Columba's death.) Máire Herbert has shown that the Irish Life was probably produced at Derry, in what is now Northern Ireland, around 1150, at the point when that monastery assumed the headship of the various interdependent foundations associated with the saint's patronal authority.[15] Herbert also points out that the Life was thus undoubtedly intended for an Irish audience, and that it is a sophisticated piece of work which continually looks to literary exemplars.

Turning to our second text, the splendidly autumnal *Acallam na Senórach* – recently translated as *Tales of the Elders of Ireland* – we find a novel-length, intricately complex collection of essentially secular tales about the great Gaelic hero Fionn mac Cumhail (Finn Mac Cool) and his associates, dated to the turn of the thirteenth century.[16] It is the longest tale extant in medieval Irish, and synthesises earlier material in a way congruent with the sophisticated tastes of high medieval audiences. As with Chaucer's *Canterbury Tales*, the *Acallam* situates multiple lesser stories within an overarching frame-tale, according to which a few straggling heroic warriors from Ireland's pre-Christian past live on mysteriously, eventually meeting St Patrick and describing their past exploits to him, as well as those of Fionn, their lord.[17] Finally, the Stowe *Táin Bó Cúailnge* is a very late medieval redaction of the greatest of all Irish sagas, the famous 'Cattle-Raid of Cooley', which exists in several versions, all zestfully ultraviolent. The story details the epic clash between the Ulstermen, with their champion Cú Chulainn, and the armies of the rest of Ireland under the formidable Queen Medb of Connaught and her husband Aillil. The version of the tale that has historically been most often translated for a general audience dates to the twelfth century, and is found in the Book of Leinster; but a starker and more

14 Further examples are discussed in the second chapter of my *Fiery Shapes: Celestial Portents and Astrology in Ireland and Wales, 700-1700*, forthcoming from Oxford University Press.

15 But see also J. Bannerman, 'Comarba Coluim Cille and the relics of Columba', *Innes Review*, (1993), Vol. 44, p. 41ff, where a case is made for Armagh as a contender for the provenance of the Life.

16 *Tales of the Elders of Ireland: A New Translation of the Acallam na Senórach*, trans. A. Dooley and H. Roe (Oxford: Oxford World's Classics, 1999).

17 See J. F. Nagy, 'Life in the Fast Lane: the *Acallam na Senórach*', in H. Fulton, ed., *Medieval Celtic Literature and Society* (Dublin: Four Courts Press, 2005), pp. 117-31.

direct eighth-century version survives in two other manuscripts, the twelfth century *Leabhar na hUidhre* and the late fourteenth century Yellow Book of Lecan, albeit in a seriously flawed form in both cases. In more recent times, it is this less flowery, earlier version that has formed the basis for the excellent translations of Thomas Kinsella and Ciarán Carson.[18] The Stowe *Táin*, however, is yet another version of the story, forming a late recension probably dating to the fifteenth century, and one which has attracted rather less critical comment.

The example of cloud-divination in the Irish Life of Columba is somewhat ambiguous, but it is likely to be the earliest example of the topos extant. The passage in question purports to describe an incident during Columba's early boyhood in the north of Ireland. When the time is coming for Columba to begin to learn to read, Cruithnechán, the boy's priestly guardian, makes suitable enquiries in the district:

> ... the priest went to a prophet (*fáith*) who was in the land to ask him when it would be right for the boy to begin. When the prophet had examined the sky, this is what he said: 'Write his alphabet for him now.'[19]

Here there seems to be no contradiction between an ecclesiastical role and consulting a secular prophet, who performs what seems likely to be cloud-divination. It is notable that it is a *fáith*, a 'prophet', who is visited here, not a *druí*, or druid. However, it is clear that the role of the *fáith* was synonymous to an extent with that of the *druí*, as the ancient linkage of the *druides* or *druidae* with the *vates* (directly cognate with Irish *fáith* and of the same meaning) suggests.

In our second example, drawn from *Acallam na Senórach*, we find the warrior Oisín, son of Fionn mac Cumhail, describing the nature of a certain hill to St Patrick. The superannuated hero's account segues into a recollection of events which happened near the hill back in the distant past. Prose gives way to six stanzas of verse in which Fionn questions his druid, Cainnelsciath, about what certain ominous clouds portend, and is duly answered:

> '"The Hill of Knowledge" is another name for it too', said Oisín. 'Why was it called that?' said St Patrick. 'Cainnelsciath, a druid of Fionn's people, used inspect the atmosphere and prophesy for Fionn, and afterwards, he spoke to him:

18 *The Tain, from the Irish epic Táin Bó Cuailnge*, trans. T. Kinsella (Oxford: Cambridge University Press, 1969), and *The Táin*, trans. C. Carson (London: Penguin, 2007).
19 *The Irish Life of Colum Cille*, in Herbert, ed. & trans., *Iona, Kells and Derry*, §21, text p. 226, trans. p. 253.

— 'Over there, Fionn', the druid said, 'you will find the hostel of Fatha Canann mac Mac Con mhic Mac Niadh mac Lughaidh. And do you see the three clouds which are over that spot?' 'Indeed, I see that' said Fionn, and he said:

Fionn: 'Cainnelsciath, over the hostel
I see three clouds brightly.
Tell everyone what the explanation for it is,
if it is allowed.'

Cainnelsciath: 'I see a cloud clear as crystal,
which is over a wide-doored hostel.
There will be a lord, if the means be strong,
The chalk of shields being shattered.

I see a cloud, grey, foreboding grief.
It is between them, in the middle;
The colour of crows and of trickeries,
A battle of weapons playing.

A crimson cloud redder than summer
I see between them up above.
From battle or from wrathful reasons
comes the colour of very red blood.

I foretell that bodies will be tormented,
the destruction early in the day of a great host;
O King of Cliu who wounds every day;
I see the three clouds of woe.'[20]

Like much medieval Irish poetry this is somewhat opaque, but it is apparent that there is a correlation between the colour of the three clouds and the prognostications that are made from them within each stanza. The bright cloud corresponds to the surfaces of the shields whitened with chalk; the grey cloud anticipates the colour of the metal weapons, *glas* being often used for the colour of blades; and the red cloud obviously prefigures the blood that will be spilled in the hostel.

Our third example of cloud-divination, that from the Stowe *Táin*, is clearer. The redactor of this version adheres fairly closely to the twelfth century Book of Leinster recension, modernising its forms in a conservative manner.[21] However, as the saga nears its end, the number of modernisations

20 For the Irish, see N. Ní Shéaghdha, ed., *Agallamh na Senórach* (Dublin: Oifig an tSolathair, 1942), pp. 156-57. For a slightly different version, see W. Stokes and E. Windisch, eds. & partial trans., *Acallamh na Senórach, Irische Texte IV* (Leipzig, 1900), pp. 211-12, (my translation).

21 See R. Thurneysen, *Die irische Helden- und Königsage bis zum 17. Jahrhundert*

increases. An episode describing cloud-divining druids led by Cathbad, *Uber*-druid of the Ulstermen, forms one of these interpolations. The hero Fergus mac Róich gives a description of one band among the assembled hosts, which consists of Cathbad along with his sons and druidic retinue:

> 'I know that man' said Fergus. 'The Foundation of Knowledge, the Master of the Elements, The Heaven-Soaring One. The Blinder of Eyes. He who takes away the strength of the enemy though the incantations of druids, namely Cathbad the fair druid, with the druids of Ulster about him; and it is for this he makes augury in judging the elements – to ascertain therefrom how the great battle ... will end. One of these clever men moreover raises his glance to heaven and scans the clouds of the sky and bears their answer to the marvellous troop that is with him. They all lift their eyes on high and watch the clouds and work their spells against the elements, so that the elements fall to warring with each other, until they discharge rain-clouds of fire downwards onto the camp and entrenchments of the men of Ireland'.[22]

The ability to manipulate the elements is the prime characteristic of druidic magic here. Again, the possibility exists that this conception is also drawn from Isidore of Seville, who explicitly tells us that the ability to agitate the elements is part of the *magus*' stock-in-trade.[23] This passage from the *Táin* in particular hints that cloud-divination is at times a kind of amalgam between the two primary roles of the druid in Irish medieval literature, combining prophecy with magical power over natural phenomena, particularly of the atmospheric kind.

Taking these texts together, we have a series of examples of cloud-divination, of which the earliest seem to be the Irish *Life of Columba* from c. 1150, which does not explicitly mention clouds, and the *Acallam* from later in the same century, which does. The scene of cloud-divination in the Stowe *Táin* is more ambiguous, and, as indicated above, could date from anywhere between the twelfth and fifteenth centuries. But the lack of literary evidence for cloud-divination from before the twelfth century suggests two possibilities. First, it is not inconceivable that the *néladóracht* topos may have existed prior to the twelfth century, perhaps extending as far back as

(Halle/Saale: Max Niemeyer, 1921), p. 117. The dating of particular interpolations within the Stowe *Táin* is difficult. On the one hand, additions to the *Book of Leinster* version might simply be inventions of the compiler of the exemplar of the Stowe *Táin*, as he undertook the process of updating and modernising. On the other, he may well have had access to other versions of the *Táin* now lost to us. In that cloud-divination seems to be a development of the twelfth and thirteenth centuries, it is very likely that the addition of the topos to the *Táin* dates to this period. See C. O'Rahilly, ed., *The Stowe Version of Táin Bó Cúailnge* (Dublin: Dublin Institute for Advanced Studies, 1978), [hereafter O'Rahilly, ed., *The Stowe Version*], pp. x-xiii.

22 O'Rahilly, ed., *The Stowe Version*, p. 142 (my translation).

23 See Isidore, *Etymologiae* viii, 9.

the Old Irish period and even beyond, as a real custom, into the pre-Christian past — but that it simply happens not to be attested in any surviving text from the early Middle Ages. (Medieval tale-lists make it dismayingly clear that a far greater corpus of Irish vernacular literature once existed than that which we now possess.) Carney's 'watchers of the heavens' would seem to imply that he took this view. The second possibility is that the topos came into existence at some point in the eleventh or twelfth centuries. I think the latter is more likely, and I think that we can link the genesis of the topos to shifts in the connotations of the word *magus* during this period.

Druids and clouds are in a sense old associates in Irish literature, as one common aspect of druidic magic in the early texts is the manipulation of fogs and mist. For example, Adomnán's *Vita Sancti Columbae*, written about the year 700, has a marvellous scene in which Columba faces down an influential Pictish *magus* (i.e., druid) called Broíchán, who conjures up a thick mist over Loch Ness to obstruct the saint's progress.[24] The specific association with atmospheric vapours is in fact one of the few elements of the early Irish image of the druid which is not directly traceable to Isidore or to the Bible. However, the magical manipulation of mists and fogs also fits well with a certain kind of standard Christian cosmology, in which the lower atmosphere is an intermediate zone where demons as well as angels are free to work. The idea is ancient, and Augustine quotes the third century Neoplatonist Porphyry of Tyre on the subject:

> In that letter, indeed, Porphyry repudiates all demons, whom he maintains to be so foolish as to be attracted by damp vapours, and therefore residing not in the ether, but in the air beneath the moon, and indeed in the moon itself.[25]

Even if the connection of druids with mists and vapours reflects pre-Christian Irish beliefs, such beliefs would be easily conflated with the demonic associations articulated by Augustine, and which Adomnán suggests were well-known in early medieval Ireland. The connotations of *magus* in the late seventh and eighth centuries were already dual, and the idea of divination as one of the activities performed by such a person could be viewed in both a positive and a negative light. On the one hand, the biblical Magi offered a positive example of pagan prophets inspired to worship Christ though their knowledge of the stars; on the other, Isidore gave a vivid and full account of wicked *magi*, incorporating many of the details of prophecy and divinatory practice found in the Old Testament and the culture of Greece and Rome.

---

24 Adomnán, *Vita Sancti Columbae*, in A.O. and M.O. Anderson, eds., *Adomnan's Life of Columba* (Edinburgh: Nelson, 1961), pp. ii, 34.

25 Augustine, *De Civitate Dei*, in J.-P. Migne, ed., *Patrologia Latina 41* (Paris, 1845), pp. x, 11.

This duality is in fact found in the New Testament itself. Matthew's Magi have skills that are pre-Christian and supernatural, yet they are neither evil nor destructive and are among the first to acknowledge Christ. Importantly, they opened the possibility of representing practitioners of astrology in a positive light, as can be seen in Tertullian's famously ambivalent statement that their art was allowable before the birth of Christ, but became illicit immediately afterwards.[26] Conversely, Simon Magus provides the New Testament paradigm for the wicked magician. Though the account of his encounter with Peter in Acts 8:9-24 is brief, he is described as a *magus* in the Vulgate and is chastised by the apostle for attempting to buy the power of the Holy Spirit. This story was greatly elaborated in the *Acts of Peter*, composed in Greek in the late second century. The majority of the text survives only in the Latin translation known as the *Vercelli Acts*, extant in a manuscript from the late sixth or early seventh century.[27] In this more elaborate legend, two contests of power occur between St Peter and Simon, in the presence of the emperor Nero; in the second, the magician demonstrates his ability to fly, but crashes to his death when Peter orders the angels who hold him up to let him fall.

As noted above, the use of the term *magus* for the class of people termed *druid* in Old Irish seems to be very old.[28] However, the equation worked both ways, so that biblical practitioners of magic could also be described in Irish with the term *druí*. It is clear that the Irish and Latin terms were effectively interchangeable, and thus the druid of early Irish literature existed within a network of semantic associations which bore absolutely no relation to native pre-Christian Celtic culture. For example, an Old Irish gloss on the *Iannes et Mambres* of 2 Timothy 3:8 — the two Egyptian sorcerers who contended with Moses and Aaron — reads *i. da druith aegeptaedi*, or 'i.e., two Egyptian druids'.[29] And the standard name for Simon Magus in Irish was *Símóin druí*, 'Simon the Druid'.

The connotations of the word *magus* in medieval Ireland were also strongly affected by the growth of a body of apocryphal legend around the biblical Magi. The Magi are in fact described notoriously vaguely in Matthew's Gospel: as the media reminds us every Christmas, it is not stated anywhere in the biblical text that there were three of them, as opposed to two, or ten. But piety, like nature, abhors a vacuum, and a legend filling in the details had reached its full form by the tenth century; this legend was well-known in Ireland from the beginning. Different racial origins, colours

26 Tertullian, in Waszink, J.H. & J.C.M. Van Winden, *Tertullianus De Idololatria* (New York: Brill, 1987), ix, 3-4.

27 For the medieval development of the legend of Simon Magus, see V. J. Flint, *The Rise of Magic in Early Medieval Europe* (Oxford: Oxford University Press, 1991), [hereafter Flint, *The Rise of Magic*], pp. 338-44.

28 Druids had been termed *magi* by Roman writers; see also McKenna, 'Between Two Worlds', pp. 67-70.

29 Stokes and Strachan, eds., *Thesaurus Paleohibernicus* (Cambridge, 1901-3), i, p. 695.

of vestments and names in dog-Greek, Latin and Hebrew were ascribed to the visitors to the Christchild — their number fixed at three — often with a symbolic interpretation attached.

The idea that the Magi had been astrologers was widely known in Ireland, an idea likely derived, as we have seen, from Isidore. Robert McNally has exhaustively studied the early Hiberno-Latin versions of the legend, and in the late eighth century text *Interrogationes vel responsiones tam de veteri quam novo testamento*, we also find the explicit designation of the Magi as astrologers:

> The question is posed, 'How did the Magi recognise the birth of Christ by means of a star?' The answer is, 'By two methods. First, from the prophecy which Balaam spoke, *a star will arise out of Jacob*; secondly, because they had been astrologers and on the Calends of January then they understood what kind of star would appear from this cause in the year of Christ's birth. It gave the Magi advance warning, so that they recognised the star with care and followed it.[30]

In several places, Hiberno-Latin references to the Magi-legend show a distinct awareness of the double nature of the divinatory power of the *magus*. In another eighth century text, which McNally argues was composed in Ireland itself, we find the following statement:

> The Chaldeans observe the motions of the stars. Elsewhere, Augustine declares that a variety of magi taught to what point of space the light of the star [of Bethlehem] was aiming in the world. These magi make their observations out of curiosity about the universe, not out of evil intent.[31]

In an eighth century Irish context, we can infer that *magi* 'of evil intent' would inevitably imply druids, recalling Muirchú's characterisations. It is telling that the author attempts to draw a distinction here between natural philosophers and malevolent magicians, reflecting an anxiety about the ambivalence of the term.[32] And in my final example from McNally, from the eighth century Irish *Liber questionum in Evangeliis*, the Magi come *ex lege, non ex stella*, a phrase which might be rendered 'licitly, not as a result of astrology'. The author is at pains to dismiss the idea that the Magi truly followed the star, preferring an allegorical and ecclesiological interpretation of the episode:

---

30 R. E. McNally, 'The Three Holy Kings in Early Irish Writing', in P. Granfield and J.A. Jungmann, eds., *Kyriakon: Festschrift Johannes Quasten* (Aschendorff: Westfalen, 1970), [hereafter McNally, 'The Three Holy Kings'], ii. p. 674.

31 McNally, 'The Three Holy Kings', p. 681.

32 Regarding the Magi as astrologers was of intermittent concern in the early medieval period, given the patristic condemnation of astrology. The rehabilitation seems to have begun in the ninth century; see Flint, *The Rise of Magic*, pp. 369-75.

> The Magi are not given a number because the faithful have been multiplied beyond number. They are not named, because the names of unbelief were given over to oblivion after the coming of the Faith, like wicked heretical *magi.*[33]

This passage goes to the absolute heart of the ambiguity of the Irish *magus*, who could equally be a noble individual integral to the coming of the Faith, or someone wickedly pagan.

We have already seen how the magicians of the Old Testament could be referred to in Old Irish as druids; the equation could even extend to Matthew's Magi themselves. In the eighth century poetry of Blathmac, the Magi are termed *na trí druídea co ndánaib*, 'the three gifted druids'.[34] And in the following quotation, the fourteenth century Leabhar Breac presents us with a version of the Magi legend in which the 'Three Holy Kings' have explicitly become druids. The legend is based on a very early Latin text of the apocryphal *Gospel according to the Hebrews*; this, together with linguistic evidence, suggests that the text itself might originally have been produced in the eleventh or twelfth centuries.[35] The text is explicitly termed 'The Tidings of the Druids' in Irish (*Scéla na nDruad*), and it runs as follows:

> On a certain day, as Joseph stood at the entrance to the house, he saw a large group approach him directly from the east. Thereupon, Joseph said to Simeon: 'Son, who are these drawing near us? They seem to have come from afar'. Then Joseph went towards them, and said to Simeon: 'It appears to me, son, that they practice druidic augury and divination, for they do not take a single step without looking upward, and they are arguing and conversing about something amongst themselves ... their appearance, colour and attire is unlike that of our own people. They are wearing bright, flowing robes, even-coloured and crimson tunics, long red cloaks and variegated gapped shoes. From their apparel they seem like kings or leaders'.[36]

33 McNally, 'The Three Holy Kings', p. 682.

34 J. Carney, ed. & trans., *The Poems of Blathmac Son of Cú Brettan* (Dublin: Irish Texts Society, 1964), p. 4.

35 The nature and language of the original text is obscure, but it must once have circulated as an independent work in a Latin version. The Irish exegete, Sedulius Scottus, writing on the continent c. 850, quotes precisely this passage in his *Collectaneum in Matthaeum*, saying that the text is known to him as the *Euangelium secundum Ebreos.* For quotation and discussion of the issues of transmission, see M. McNamara, 'Apocryphal Infancy Narratives', in P. Ní Chatháin and M. Richter, trans., *Ireland and Europe in the Early Middle Ages: Texts and Transmissions/Irland und Europa im früheren Mittelalter: Bildung und Literatur* (Dublin: Four Courts Press, 2002), pp. 127-29.

36 M. Herbert and M. McNamara, eds. & trans., *Irish Biblical Apocrypha: Selected Texts in Translation* (Edinburgh: T. & T. Clark, 1989), p. 36.

This passage is strongly reminiscent of the description from afar of Cathbad and his marvelous entourage from the Stowe version of the *Táin*, with the important difference that *magic* is absent here: both groups observe the heavens for the purposes of augury, but these Magi do not manipulate the elements. This passage in fact gives us the crux of the relationship between cloud-divination and astrology. Looked at in the context of the vernacular saga literature, the form of augury the 'druids' practice here would naturally be taken to be cloud-divination. But viewed in terms of the medieval legend of the Magi, it would be equally natural to take it as astrology. A kind of discreet elision has occurred, and the actual nature of the 'druidic augury' has been left ambiguous.

We have surveyed a wide range of evidence in the course of this article. What conclusions can we draw, therefore, about the phenomenon of cloud-divination in medieval Irish literature? Ultimately, the creation of the topos seems likely to date to the eleventh or twelfth century, and to be an unexpected consequence of interchangeability of the terms *magus* and *druí* in Irish medieval culture. It is unlikely, I think, to be a historical druidic custom inherited from the pagan past (if it were, one would surely expect it to be attested earlier that the mid-twelfth century, somewhere among all the vivid images of prophesying druids which are extant in early Irish texts.)

Further, it is plain that the slippage of the connotations of *magus* and *druí* was both complex and dynamic over the centuries. From the earliest, literary druids were traditionally invested with the power of prophecy, one of the few aspects of their characterisation which seems likely to reflect the activities of their historical flesh-and-blood counterparts. The star-gazing of our earliest heaven-scanning druid, in the *Vita Prima* of Brigit, seems to have been influenced by Isidore's highly authoritative description of *magi* — and *the* Magi — as astrologers, in the middle of the seventh century. That druid's ambiguous position as both a pagan wizard (bad) and as a solitary analogue to the Magi present at Christ's birth (good), underscores the ambivalence which the term *magus* carried in Ireland from the beginnings of literary activity there.

The druids' traditional association with clouds and mist dovetailed well with a Christian cosmology in which the lower atmosphere was seen as demon-haunted, and thus as particularly amenable to manipulation by means of malevolent pagan magic. But, on the other hand, the Magi of the New Testament provided a useful model when Irish literati wanted to create more sympathetic images of druids, as native prophets illuminated by a certain amount of natural grace. (The druid in the *Vita Prima* prophetically recognises Brigit's sanctity, for example, but she is unable to keep down the food he provides for her, as he is, in some sense, *unclean* by virtue of his paganism.) So these, as it were, are our ingredients for the creation of *néladóracht.*

Two developments coincided with the gap between the star-gazing *magus* of the *Vita Prima* in the mid seventh century, and the earliest episodes of cloud-divination in the mid-to-late twelfth century. The first was the full elaboration of the medieval legend of the Magi, which, as we have seen, the Irish knew as well as anyone; and second was a general relaxation of ecclesiastical anxieties about astrology, presaging its widespread revival in the twelfth century. Such conditions seem to have allowed the transference of the locus of divination from the stars to the clouds, perhaps as a naturalisation, undoubtedly influenced by the ancient association of druids with mists, long hallowed by the oldest layer of Irish hagiography.

As a further observation on this point, it is notable that the word for 'soothsayers' in Isaiah 2:6 and Micah 5:11, rendered *augures* in and *divinationes* respectively in the Vulgate, is in the Hebrew text literally 'cloud-observers'.[37] Micah in particular is explicit in underlining divine condemnation of such people, in the context of a furious prophecy of coming destruction for the heathen, along with their witchcraft, graven idols and sacred groves: ... *et auferam maleficia de manu tua, et divinationes non erunt in te*, or, as the Authorised Version vigorously renders it, 'And I will cut off witchcrafts out of thine hand; and thou shalt have no more soothsayers'. Though this is a prophecy of future destruction, learned Irish churchmen might very well have taken it allegorically as a reference to the past of their own island, where they themselves were once the heathen, with native priest-prophets who apparently worshipped idols in sacred groves.[38] It is just possible that they were aware of the meaning of the term in Hebrew: certainly some Hebrew at least was known in early medieval Ireland.[39] But if creating a new literary form of divination was intended to distinguish native druidic *magi* from the biblical ones, it failed dismally, as the semantic overlap seems to have worked both ways. The marvellous troop of Magi described in *Scéla na nDruad*, would, one feels, be just as at home on the battlefields of Ulster as in the biblical stable. And so it seems that at the start of Irish literary activity, druids were *magi*, and inherited all the negative Patristic associations to which such personages were heir; but by the end of the Middle Ages in Ireland, and probably long before, the Magi themselves had become druids. And the art of astrology, which may well

37 The Hebrew word is *anan*, which means 'to practice soothsaying', though the etymological root is 'cloud'. See *The Brown, Driver, and Briggs Hebrew and English Lexicon*, ed. F. Brown (Peabody, Massachusetts: Hendrickson Publishers, [1906, 2001,] 2004), pp. 777-78. Note also that *Young's Analytical Concordance to the Bible*, ed. R. Young (Nashville: Hendrickson Publishers, 1982), p. 915, gives 'to observe the clouds' as the literal meaning for the passages in Isaiah and Micah. See also A. Jeffers, *Magic and Divination in Ancient Palestine and Syria* (Leiden: Brill, 1996), pp. 78-81, for background on the possible practice of *anan*. I am grateful to Helen Jacobus of the University of Manchester for this insight.

38 This would be far from an unusual strategy. See McCone, *Pagan Past*, pp. 32-7.

39 See, for example, the work of my colleague Pádraic Moran, 'Sacred languages and Irish glossaries: evidence for the study of Latin, Greek and Hebrew in early medieval Ireland' (unpublished PhD thesis, National University of Ireland, Galway, 2007).

have inspired the unusual motif of cloud-divination, had become obscurely and interchangeably confused with it.

Bibliography

Adomnán, *Vita S. Columbae*, in A.O. and M.O. Anderson, eds., *Adomnan's Life of Columba* (Edinburgh: Nelson, 1961).

Adomnán, *Life of St Columba*, trans. R. Sharpe (Harmondsworth: Penguin, 1995).

Augustine, *De Civitate Dei*, in J.-P. Migne, ed., *Patrologia Latina* 41 (Paris, 1845).

Bannerman, J., 'Comarba Coluim Cille and the relics of Columba', *Innes Review,* (1993), Vol. 44, pp. 14-47.

The Brown, Driver, and Briggs Hebrew and English Lexicon ed. F. Brown (Peabody, Massachusetts: Hendrickson Publishers, [1906, 2001,] 2004).

Caesar, *Comentarii de Bello Gallico*, in H.J. Edwards, ed. & trans., *The Gallic War* (Cambridge, Massachusetts: Loeb Classical Library, 1917).

Carney, J., 'Language and literature to 1169', in D. Ó Cróinín, ed., *A New History of Ireland, Volume I: Prehistoric and Early Ireland* (Oxford: Oxford University Press, 2005), pp. 451-509.

Carney, J., ed. & trans., *The Poems of Blathmac Son of Cú Brettan* (Dublin: Irish Texts Society, 1964).

Charles-Edwards, T.M., *Early Christian Ireland* (Cambridge: Cambridge University Press, 2000).

Flint, V.J., *The Rise of Magic in Early Medieval Europe* (Oxford: Oxford University Press, 1991).

Hutton, R., *Blood and Mistletoe: the History of the Druids in Britain* (New Haven & London: Yale University Press, 2009).

Irish Biblical Apocrypha: Selected Texts in Translation, trans. M. Herbert and R. McNamara (Edinburgh: T. & T. Clark, 1989).

The Irish Life of Colum Cille, in M. Herbert, (ed. & trans.), *Iona, Kells and Derry, The History and Hagiography of the Monastic Familia of Columba* (Oxford: Oxford University Press, 1988), pp. 218-65.

The Irish Penitentials, ed. L. Bieler (Dublin: Dublin Institute for Advanced Studies, 1975).

Isidore of Seville, *Etymologiae*, in W. M Lindsay, ed., *Isidori Hispalensis Episcopi Etymologiarum Sive Originum Libri XX*, 2 Volumes (Oxford: Oxford University Press, 1911).

Jeffers, A., *Magic and Divination in Ancient Palestine and Syria* (Leiden: Brill, 1996).

Kelly, F., *A Guide to Early Irish Law* (Dublin: Dublin Institute for Advanced Studies, 1988).

McCone, K., *Pagan Past and Christian Present in Early Irish Literature* (Maynooth: An Sagart, 2000).

McKenna, C., 'Between Two Worlds: Saint Brigit and Pre-Christian Religion in the *Vita Prima*', in J.F. Nagy, ed., *Identifying the 'Celtic': the Celtic Studies Association of North America Yearbook 2* (Dublin: Four Courts Press, 2002), pp. 66-74.

McNally, R.E., 'The Three Holy Kings in Early Irish Writing', in P. Granfield and J.A. Jungmann, eds., *Kyriakon: Festschrift Johannes Quasten* (Aschendorff: Westfalen, 1970).

McNamara, M.,'Apocryphal Infancy Narratives', in P. Ní Chatháin & M. Richter, eds., *Ireland and Europe in the Early Middle Ages: Texts and Transmissions/Irland und Europa im früheren Mittelalter: Bildung und Literatur* (Dublin: Four Courts Press, 2002), pp. 123-46.

Moran, P., 'Sacred Languages and Irish Glossaries: Evidence for the Study of Latin, Greek and Hebrew in Early Medieval Ireland' (unpublished PhD thesis, National University of Ireland, Galway, 2007).

Nagy, J.F., 'Life in the Fast Lane: the *Acallam na Senórach*', in H. Fulton, ed., *Medieval Celtic Literature and Society* (Dublin: Four Courts Press, 2005), pp. 117-31.

Nash, D., 'Reconstructing Posidonius's Celtic Ethnography', *Britannia,* (1976), Vol. 7, pp. 112-36.

Ní Shéaghdha, N., ed., *Agallamh na Senórach* (Dublin: Oifig an tSolathair, 1942).

Ó Cróinín, D., 'Ireland, 400-800' in D. Ó Cróinín, ed., *A New History of Ireland 1: Prehistoric and Early Ireland* (Oxford: Oxford University Press, 2005), pp. 182-233.

O'Rahilly, C., ed., *The Stowe Version of Táin Bó Cúailnge* (Dublin: Dublin Institute for Advanced Studies, 1978).

Stokes, W. and J. Strachan, eds., *Thesaurus Paleohibernicus,* 3 Volumes (Cambridge, 1901-3).

Stokes, W. and E. Windisch, eds. & partial trans., *Acallamh na Senórach, Irische Texte IV* (Leipzig, 1900).

Waszink, J.H. & J.C.M. Van Winden, *Tertullianus De Idololatria* (New York: E.J.Brill, Leiden, 1987).

The Tain, from the Irish epic Táin Bó Cuailnge, trans. T. Kinsella (Oxford: Cambridge University Press, 1969).

The Táin, trans. C. Carson (London: Penguin, 2007).

Tales of the Elders of Ireland: A New Translation of the Acallam na Senórach, trans. A. Dooley and H. Roe (Oxford: Oxford University Press, 1999).

Thurneysen, R., *Die irische Helden- und Königsage bis zum 17. Jahrhundert* (Halle/Saale: Max Niemeyer, 1921).

Vita Prima S. Brigitae, in J. Colgan, ed., *Triadis Thaumaturgae ... Acta* (Louvain, 1647), pp. 527-42.

Young, R., ed., *Young's Analytical Concordance to the Bible* (Nashville, Tenessee: Hendrickson Publishers, 1982).

Astrology in the Seventeenth-Century Scottish Universities

Jane Ridder-Patrick

There is evidence that astrological concepts were being taught, or engaged with, in all four Scottish universities until at least 1700. For that reason alone, astrology has a valid claim to be considered a component of contemporaneous mainstream intellectual life. This paper will address the questions of what — and who — shaped the university curricula and those men's attitude toward astrology. It will examine the manuscript evidence that demonstrates how, where and why astrology could fit into a predominately Presbyterian educational system, the factors that sustained its presence until this relatively late date and those that facilitated its demise.

While detailed studies have been undertaken of astrology in England and many areas of continental Europe, its importance for the thought and culture of early modern Scotland has, until now, been entirely neglected. This paper outlines for the first time astrology's place in Scottish academic history and demonstrates that astrological theory was taught at the four ancient universities — St Andrews, Aberdeen, Glasgow[1] and Edinburgh — until the final quarter of the seventeenth century. It argues that this was facilitated by support given to astrology by influential figures, resistance to debate on natural philosophy by certain clergy and by the regenting system of instruction. It shows too that, despite an apparent upsurge in interest in the 1670s, academic attitudes towards astrology had shifted radically by the end of the century and contends that factors largely contributing to this include changes in the syllabus and teaching methods which accompanied key developments in the mathematical sciences.

The main primary sources for this investigation are student lecture notes found in the libraries of those four universities and in the National Library of Scotland. These are supplemented by university library catalogues which give data about book holdings, donations and purchases. So far I have identified sixteen notebooks that contain astrological material of one kind or another, and each of the four universities is represented. The earliest exemplar is dated 1613-14 and the latest is from 1700.

---

1 The material from the Glasgow notebooks is still at the transcription and translation stage and, although astrological material is apparent in these, it has not been included in this study.

In what follows, the nature of the astrological material that was being taught or discussed, and where it fitted into the academic curriculum, will be examined, and in each of the notebooks an attempt made to distinguish which of the two branches of astrology was being dealt with — *natural* or *judicial.*

Natural astrology links the movement of the heavenly bodies with natural phenomena such as the weather, agriculture and medical matters and this practice was largely accepted as self-evident. Judicial astrology, by far the more controversial branch, covers general predictions, the election of the best times for starting projects, nativities, decumbitures and horary questions. The four foundational components of astrology are the planets, signs, aspects and houses. The first three are used in both natural and judicial astrology. In general the houses — when they refer to the different areas of life affected by the planetary movements — are used only in judicial astrology. The presence in the notebooks of the meanings of the houses, as well as zodiacal divisions known as terms, will be used as an indication that judicial astrology as well as natural astrology was being covered. Further, the questions of why astrology's presence was condoned as part of Scottish university education in the early years of the seventeenth century, yet condemned by the beginning of the eighteenth century, and when and why attitudes towards it changed, will be touched upon.

Teaching at all of the Scottish universities until the eighteenth century was by the regenting system. In this system, the same teacher, known as a regent, instructed a class on each subject in turn, all the way through the four years of the Arts course.[2] There were lectures in the mornings, followed in the afternoon by disputations on what had just been taught, so that through debate the students could absorb the lessons more thoroughly. The regents read out their lectures in Latin at dictation speed to the students who wrote them down more or less word for word in their notebooks. The notebooks generally contain the material taught over the course of one academic year and they give a good account of the dates of the lessons, the subject matter taught and even the exact words that the regent had used. This provides the researcher with a superb record of the details of what was being taught on an almost daily basis. Unfortunately, particulars of the disputations have rarely been documented.

Students normally began their studies around the age of fourteen when they entered the Bajan or first class. Until the 1660s the teaching material, after the Bajan year when Greek was taught, consisted mainly of commentaries on the works of Aristotle. In general, Logic and Metaphysics were introduced in the second year, the third year covered Ethics and more Metaphysics, and the fourth or magistrand year was taken up with Physics

2 Andrew Dalzel, *History of the University of Edinburgh, Volume 2* (Edinburgh: Edmonston & Douglas, 1862).

and Natural Philosophy.[3] It is in the astronomy, physics or special physics sections of the magistrand-year student notebooks that all of the astrological material so far has been found. The vehicles for its introduction in the earlier years of the seventeenth century were mainly either the commentaries on Sacrobosco's *Sphere,* or Book II of Aristotle's *De Caelo*, or both. An examination of five of the notebooks from 1613, 1660, 1662, 1670 and 1672 shows the range of material being taught.

Astrology appears twice in the lectures of the regent James Reid which were taken down at Edinburgh in 1613 by Alexander Henryson.[4] The first is in a lecture on annotations on chapter two of Sacrobosco's *Sphere* in which Reid gives instruction on the planetary strengths and weaknesses, naming the signs in which each of the planets has its house, detriment, exaltation and fall. This is followed by a table providing an at-a-glance overview of what has been taught but there is no commentary on how these are to be used. A discussion of its applications — and given the baldness of the factual presentation it is difficult to imagine that one did not follow it — would have taken place in the afternoon disputations.

The second astrological entry is a printed image of a Zodiac Man, pasted into the book, which had almost certainly been cut out of an old almanac. This depicts a naked man with the creatures of the zodiac around him, showing the links between the signs and various parts of the body. The accompanying text, written by Henryson, goes along, more or less, with the traditional correspondences.[5] None of this can be said with any certainty to go beyond the bounds of natural astrology.

The next extant notebook, dated 1660, is from St Andrews. It belonged to Colin Campbell, a student who went on to become a remarkable minister,

3 Christine Shepherd, *Philosophy and Science in the Arts Curriculum of the Scottish Universities in the Seventeenth Century* (PhD Thesis, Edinburgh University, 1974).

4 National Library of Scotland, Adv.Ms.5.2.3, f.125v, 126r.

5 Henryson writes (in National Library of Scotland, Adv.Ms.5.2.3, f.171r.):
In order to predict from the heavens there are the twelve signs
The man's body is assigned thus from the top
For instance the head and face themselves rejoice in Aries
The throat and neck are where Taurus dominates
The arms and hands are supported by Gemini
It is the nature of Cancer to stretch forth her breast
And Leo wishes the stomach for itself
But the intestines Virgo desires above all
Libra has dominion over both buttocks
Scorpio wishes to have the shameful parts
Faithful Sagittarius wants to rule the hips
Capricorn has power over both knees
Finally the calves are suitably declared for Aquarius
And Pisces is the owner from [the] [ankles?] to the soles of the feet.

mathematician and astrologer.[6] The astrological section appears in disputations in Book Two of Aristotle's *de Caelo,* and begins:

> Astrologers boasted that it is possible to predict all of the future that will come forth into being because of celestial influxes from various astrological aspects, and furthermore most contingently.[7]

The notes go on to give a description of the Ptolemaic planetary aspects and this is illustrated by a diagram showing how they are connected within the circle that represents the sky. There is also a section on the planets, signs and elements, while a second illustration shows how astrologers divide the sky at any particular moment into the twelve houses of the horoscope. While it could not be said with any certainty that Reid's dictates ventured outside the bounds of natural astrology, this does step squarely into the territory of predictive and judicial astrology.

George Gordon (1637–1720), who later became the first Earl of Aberdeen,[8] was a gifted scholar who graduated from King's College at the head of his class in 1658 under the regent Andrew Massie. Difficult political times forced him to become a regent himself, instructing one class all the way through its four years on the course that he had himself only just completed. As he taught astrology, it is a fair assumption that his regent, Andrew Massie, did so too. More will be said of Massie later.[9] A notebook of Gordon's 1662-3 magistrand class has survived, belonging to John Barclay, son of Lady Johnstone.[10] Astrological notes are found both in the commentaries on Aristotle's *De Caelo*, and under the general heading of astronomy. There is a diagram of an astrological figure, divided into the twelve houses, each of which represents a different area of life and each house in turn is ascribed a meaning as follows:

> 1.*Vita*, 2. *lucrum*, 3. *fratres*, 4. *genitor*, 5. *nati*, 6. *valetudo*, 7. *uxor*, 8. *mors*, 9. *religio*, 10. *regum/regna*, 11. *benefactor*, 12. *carcer.*[11]

---

6 EUL-Ms.3101.6. Colin Campbell's regent was William Campbell.

7 EUL-MS.3101.6, fol. 256.

8 J. D. Ford, 'Gordon, George, First Earl of Aberdeen (1637–1720)', in the *Oxford Dictionary of National Biography* (Oxford: Oxford University Press, 2004).

9 By the time that Gordon had completed teaching his class, the political situation had become more favourable for him and he was able to leave to study abroad. When he was elevated to the peerage in 1682 he chose for his motto *Fortuna sequatur* ('let fortune follow'), a sideward glance perhaps to his astrological studies.

10 National Library of Scotland, Adv.Ms.22.7.15.

11 These can be translated as: 1. Life, 2. Money, 3. Siblings/cousins, 4. Father, 5. Children, 6. Health, 7. Wife, 8. Death, 9. Religion, 10. The king/power, 11. Benefactor, 12. Prison. From National Library of Scotland, Adv.Ms.22.7.15, fol. 208.

Later in the notebook there is a diagram of the planetary aspects — conjunctions, sextiles, squares, trines, and oppositions — and also a table of the planetary terms. These show the degrees of the zodiac in which each planet has special influence. Significantly, terms are only used in judicial astrology to help weigh up and judge the likely outcome of questions put to the astrologer.

An Aberdeen notebook from 1670 of lectures given by an unidentified regent contained similar material on the aspects and houses. This regent uses slightly different terminology for the interpretation for some of the houses. For example, the eleventh and twelfth houses now become *Bonus daemon* and *Cacodaemon*.[12] The significance remains the same but the variation suggests an alternative source of information. Also included are the dignities and debilities of the planets, and on the final page of the notebook data is given for the nativity of Charles II (1630–1685). In these two Aberdeen notebooks, the line between judicial and natural astrology has been quite definitively crossed.

The latest-dated notebooks found containing astrological teaching are in the lectures of the Edinburgh regent James Pillans. Two survive from his 1672 magistrand year.[13] Although Aristotle's second book of *De Caelo* is dealt with, the astrological material comes as commentaries on Sacrobosco's *Sphere* under the heading of 'de Zodiaco'.[14] Pillans, like Reid in 1613, gives the rulerships of the planets and a table of planetary dignities and debilities and there is likewise no information on houses. In the notebooks found so far, the astrological material in those from Aberdeen and St Andrews is overtly judicial, while in those from Edinburgh it is not. Further evidence must be examined before the question can be answered satisfactorily of whether this is simply an artefact of the sample or an indication that at Edinburgh, unlike the other universities, judicial astrology was censored while natural astrology was sanctioned.

It is virtually certain that the teaching of this material was non-controversial, as the universities were heavily regulated, with the Kirk and university authorities — and in the first quarter of the century King James VI himself — taking a keen and interfering interest in the curriculum that was to be taught, and carrying out periodic visitations to suggest reforms. In Edinburgh there was an extra layer of control as the town council governed the university and it even laid down the specifics of what was to be taught, as for example in the 'Fourt Yeir':

12 His complete list can be traslated as: 1. Life, 2. Prospects, 3. Goddess and brothers, 4. [he misses out the meaning for the fourth house], 5. Children, 6. Health, 7. Spouse, 8. Death, 9. Religion, M.C. [no house number given], 11. Good daemon, 12. Bad daemon. From National Library of Scotland, Acc.4975, fol. 140.

13 EUL-Gen.2028 and EUL-Dc. 6.5.

14 Explicatio doctrinae Astronomicae ex Spher Joannis a Sacrabosco, p. 82ff.

> The examinatiounes being endit the Regent teacheth unto theme the first sum pairt of the second and the fourt buik De Caelo. The buikes De Caelo being endit he teacheth Sphaeram Joannis de Sacra Bosco, and out of his fourt chapter gives theme some insicht in the theoretick of the planets magnitudes and constellatiouns of the starris. The sphere being endit he teacheth the buikes De Ortu et Interitu and the Metiores so far as is neidfull.[15]

The Town Council also decreed that the student notebooks had to be examined once a year to ensure they were up to standard. As these dictates contained word-for-word records of what was taught in the lecture rooms, it is most unlikely that a regent would risk losing a highly sought-after post by inserting contraband material. It is possible that it was these examinations, or the threat of them, that made the Edinburgh regents more wary of the explicit teaching of judicial astrology.

Calvin, the inspiration for the Scottish Reformation, had written a tract against judicial astrology.[16] So why then was astrology of both kinds condoned in the universities of Scotland for such a prolonged period? It may have been in part that intellectual stagnation allowed ancient practices to continue in the three oldest universities. But Edinburgh University, the newest of the traditional universities, was founded after the Reformation in 1583.

One possible reason is that a number the Scottish intellectual elite in the late sixteenth and early seventeenth century were keen astrologers, astronomers and mathematicians. This group of friends included John Napier, the inventor of logarithms, and the influential minister and judge Robert Pont, one of John Knox's right-hand men. Pont was six times moderator of the Church of Scotland and an advisor to James VI. His attitude to astrology was that:

> many evident signes are founde in the motiones, configurations, and interchangeings of the courses of the heavenly light, whereby men, who are expert in divine science of Astrology, may gather and conjecture, many things to fall out, not only in the aire, but also in the naturall inclination of earthly creatures.

Moreover,

> For that cause, the eternall God appoynted them in the beginning, not onely to shine and shew lights vnto the world, but also to be for signs of things to come: as it is testified in Genesis.[17]

---

15 Alexander Morgan, ed., *University of Edinburgh: Charters, Statutes, and Acts of the Town Council and the Senatus, 1583-1858* (Edinburgh: Oliver & Boyd, 1937), p. 113.

16 John Calvin, *An Admonicion against Astrology Iudiciall*, trans. Goddred Gilby (London, 1561).

17 Robert Pont, *A Newe Treatise* (Edinburgh, 1599), p. 45.

Astrology therefore, according to Pont, was sanctioned by God. As the first Bible to be printed in Scotland was going to press in 1579 Pont petitioned the General Assembly of the Church of Scotland to have his calendar and almanac inserted into it, and permission was granted for this. Part of this was a table for finding the sign and degree of the Moon at any time and included the qualities of the signs — temperate and cold and moist and dry. The only feasible explanation for the inclusion of this would have been to provide information that was essential for the practice of natural astrology. There were other supporters of astrology in high places too. Just one example is the king's physician in Scotland, George Eglisham, who had a highly detailed astrological workbook printed in 1616.[18]

By the 1680s the situation had changed, certainly in the universities, and for the next 20 years all mention of astrology in the extant notebooks is of a condemnatory nature. As the surviving notebooks are from only a fraction of the total classes given in all of the universities, care must be taken in drawing conclusions, but a shift in attitude seems to have taken place in the 1670s. The 1682 dictates of the regent Andrew Massie[19] who had just moved from Aberdeen to Edinburgh contained six pages of objections to judicial astrology. Condemnation of astrology is found in magistrand year notebooks over the following 20 years until at least 1700, a sure indication that academics perceived there to be a threat of some sort from astrology. There is no reason otherwise why they would have wasted time and attention on it. The objections, stripped to their basics, tend to come back to accusations of impiety, charlatanism, poor methodology and inaccuracy of predictions. So what had happened to about bring this shift?

The answer may lie partially in gradual changes in the regenting system which had hardly encouraged innovation and specialisation. Regents tended to be appointed from the best of the newly graduated students, and as there would rarely be opportunity for wider experience between graduation and beginning to teach, the views and methods of the regent's own teacher tended to be perpetuated, sometimes word for word. James Pillans was taught by Robert Rankin who in turn was taught by James Reid. Pillan's astrological notes of 1672 are virtually identical to those of Reid in 1613. There was also little time available for personal research. Regents worked long hours with classes beginning at 5am in summer and 6am in winter.[20] In addition to these constraints, the there were those in the powerfully influential Kirk who opposed to developments in natural philosophy and exerted a stifling effect on innovation and creativity. One determined churchman forced the dismissal of the aforementioned Reid in 1627 for championing liberal science, despite the fact that the Town Council wanted to retain him as he was an excellent teacher. Paradoxically, this kind of

18 George Eglisham, *Accurata Methodus* (Edinburgh, 1616).

19 EUL-Dc.6.23.

20 This was put back an hour later in the century.

repression may well have had the effect of keeping astrology in the curriculum, as part of the established body of knowledge, for longer than it might have been otherwise.[21]

During eleven years, from 1667 to 1682 at Edinburgh University the same four regents — Pillans, Wishart, Wood and Paterson — had a virtual monopoly of teaching. They taught all of the students and they taught the traditional material in the traditional manner. An examination of the Edinburgh University library purchases shows another interesting connection. All of them were associated with the purchase of astrology books. Between 1670 and 1679 they bought or acquired the following: William Lilly's *Introduction to Astrology,* Nostradamus' *Prophecies,* Gadbury's *Collection of Nativities & observations historical thereupon,* Coley's *Key to Astrology* and J B Morini's *Astrologia Gallica.* These are all key texts in judicial astrology. But *Gaule against Astrology alias his Magastromancer,* a popular attack on astrology, was also purchased in 1674 and in 1677 another copy of this was handed in.[22] As there were very few astrology books acquired between the beginning of the seventeenth century and the 1670s these purchases suggest a surge in interest in and engagement with astrology at that time. While the increased output and availability of astrology books from England may account in part for this, action does not of necessity follow opportunity, unless other factors are at work. Astrology until the 1670s appears to have been embedded in traditional texts, ossified into tradition, an accepted but unreflected-upon part of the canon, but now it seems the debate was on. Astrology was fighting for its place in mainstream academic thinking.

At this time the grip of the regenting system was beginning to loosen with the introduction of professorial posts. In 1674 the mathematician and astronomer James Gregory was invited by Edinburgh University to take up the newly established chair of mathematics.[23] This allowed for more specialisation and innovation as opposed to the repetition of the same lectures year after year. Concurrently, especially in Edinburgh, changes were being introduced in the course material of the magistrand year. Whereas previously the bias had been heavily Aristotelian, with the astronomical theories of Ptolemy, Tycho Brahe and Copernicus all offered as explanations of planetary phenomena, usually without a preference being expressed; Aristotle was now being replaced with such topics as chronology, geography and pure astronomy, and Cartesian theory was being introduced.

21 There were, as might be expected, personal and political factors involved in Reid's dismissal, but there was very little change in the teaching of astronomy and natural philosophy after that date until the late 1670s.

22 EUL-Da.1.29-33.

23 Unfortunately, in October 1675, a few months after his arrival in Edinburgh, a stroke blinded him while he was showing Jupiter's satellites to his students and he died a few days later.

It was often in the Cartesian context that astrology, specifically judicial astrology, was criticised.

As Aristotle and Sacrobosco were dropped from the syllabus, astrology's traditional platform went too. Astrology is not theoretically or doctrinally supported by Christianity, although a forcible argument can be, and was, made either for or against it using scripture. It is however supported to a large extent by the theories of Aristotle and Sacrobosco.[24] It is indicative of the untenability of astrology's position at the time that another means of introducing it into the syllabus was not found. Instead the symbolic worldview of astrology was finally displaced and replaced by the kind of mathematical certainties epitomised by Newton's *Mathematical Principles of Natural Philosophy*. These laws produce verifiable predictions about the material world while astrology, especially judicial astrology with its emphasis on subjective interpretation, does not lend itself readily to such direct verification. Astrology had been an important tool, arguably the most important one, in attempting to comprehend and manipulate the influences that controlled life on earth. Now it was up against a superior contender for this, and one that held the key not only to understanding the natural world but to economic advantage. Whatever the merits of traditional astrology, its flame could burn but dimly in contrast to the beacon of this new light.

In conclusion it can be seen that until the 1670s instruction in astrology was given at all four Scottish universities. Judicial astrology was included in lectures at Aberdeen and St Andrews, while at Edinburgh[25] it remains unclear whether anything other than natural astrology was taught. After an upsurge of interest in the subject in the 1670s, during the 1680s and 1690s judicial astrology came under attack at all of the universities and by the early years of the eighteenth century, concurrent with a shift from the regenting to the professorial system and changes in the magistrand syllabus, astrology had disappeared from the academic horizon.

---

24 The following is from the English translation of Sacrobosco's Sphere in Lynn Thorndike, *The Sphere of Sacrobosco and its Commentators* (Chicago: University of Chicago Press, 1949), p. 124: 'That circle is called "zodiac" from *zoe*, meaning "life," because all life in inferior things depends on the movement of the planets beneath it ... By Aristotle in *On Generation and Corruption* it is called the "oblique circle," where he says that, according to the access and recess of the sun in the oblique circle, are produced generations and corruptions in things below'.

25 And possibly Glasgow.

Bibliography

Primary Sources

National Library of Scotland
Adv.MS.5.2.3, f.125v, 126r.
Adv.Ms.22.7.15.

University of Edinburgh Library
Ms.3101.6.
Gen.2028.
Dc.6.5.
Dc.6.23.
Da.1.29-33.

Printed Primary Sources

Calvin, John, *An Admonicion against Astrology Iudiciall*, trans. Goddred Gilby (London, 1561).
Eglisham, George, *Accurata Methodus* (Edinburgh, 1616).
Pont, Robert, *A Newe Treatise* (Edinburgh, 1599).

Secondary Sources

Cant, Ronald Gordon, *The University of St Andrews: A Short History* (Edinburgh: Oliver & Boyd, 1970).
Dalzel, Andrew, *History of the University of Edinburgh, Volume 2* (Edinburgh: Edmonston & Douglas, 1862).
Donaldson, Gordon, ed., *Four Centuries: Edinburgh University Life, 1583-1983* (Edinburgh: Edinburgh University Press, 1983).
Ford, J. D., 'Gordon, George, First Earl of Aberdeen, 1637–1720', *Oxford Dictionary of National Biography* (Oxford: Oxford University Press, 2004), http://www.oxforddnb.com/index/101011039/George-Gordon [accessed 14 October 2009].
Morgan, Alexander, ed., *University of Edinburgh: Charters, Statutes, and Acts of the Town Council and the Senatus, 1583-1858* (Edinburgh: Oliver & Boyd, 1937).
Shepherd, Christine, *Philosophy and Science in the Arts Curriculum of the Scottish Universities in the Seventeenth Century* (PhD Thesis, Edinburgh University, 1974).
Stevenson, David, *King's College, Aberdeen, 1560-1641: from Protestant Reformation to Covenanting Revolution* (Aberdeen: University of Aberdeen Press, 1990).
Thorndike, Lynn, *The Sphere of Sacrobosco and its Commentators* (Chicago: University of Chicago Press, 1949).
A Catalogue of the Graduates in the Faculties of Arts, Divinity, and Law, of the University of Edinburgh since its Foundation (Edinburgh: Bannatyne Club, 1858).

Decoding the Inter-Textual Literary Strata of the Mummers' Play: Some Unexpected Astronomical Themes and a Pagan 'Fingerprint' – Continuity or Reconstruction?

Glenford Bishop

The Mummers' Play, or English folk play as folklorists refer to it, was for many years thought to be a relic of pagan survivals. This perspective was finally abandoned when the evolutionary survivals theory became obsolete. There is no alternative theory. However, analysis of the texts through a structural hermeneutical literary approach, informed by the work of Lacan and Riffaterre, resulted in a reconciliation of the pagan origins theories with the contemporary folklorists' perspectives of eighteenth century invention. This analysis penetrates through the textual layers of the plays and reveals astronomical and other themes derived from, and located in, the eighteenth century's field of Celtic reconstruction.

The Mummers' Play

The Mummers' Play forms part of a wider *genre* known collectively as The English Folk Play, comprising the Sword Dance Play and Plough Play as well as the Hero-Combat Play, which is the folklorist's term for the Mummers' Play.[1] This paper is focusing on the latter only.

The terms 'Mummer' and 'Mumming' have been used consistently in folklore records referring to masking, disguising and sometimes cross-dressing customs, which were seasonal, chiefly around Christmastime in many parts of the British Isles. The purpose seems to have been to bring good luck to those visited in exchange for money, alcoholic drinks and seasonal foodstuffs.[2] Michael Preston proposed a theory that these 'luck

1 See Edmund Chambers, *The English Folk Play* (Oxford: Clarendon Press, 1933), [hereafter Chambers, *English Folk Play*], for classification of the plays, and also Peter Millington, 'The Origins and Development of English Folk Plays' (PhD Thesis, Sheffield University, 2002), [hereafter Millington, 'Origins and Development'], where this classification is further discussed, and the term Mummers' Play clarified.

2 See Jacqueline Simpson and Steve Roud, *Dictionary of English Folklore* (Oxford: Oxford University Press, 2003), p. 250.

visiting' customs were elaborated sometime in the eighteenth century, when records of extended luck visits incorporating the Mummers' Play were first recorded, a theory which is now generally accepted amongst folklore scholars.[4]

These early plays, appearing at manor houses, marketplaces and village streets, involved two principal characters: St George and Alexander the Turkish Knight. They enter into a blustering combat wielding their wooden swords. When one of them falls to the ground, seemingly dead (and this varies from play to play and district to district as to which character is killed), there is a call for a doctor. The Doctor steps forward with his medicinal bag, curing the fallen one with a quack remedy, usually announced as the herb Elecampane. Following this seemingly miraculous resurrection, walk-on characters, such as Beelzebub, Big Head, or Little Johnny Jack, step forward with bowls, boxes or other vessels, claiming money, beer, food, and usually ending the performance with a song or dance.[5]

For much of the past two centuries, folklorists have been convinced that the Mummers' Plays are the vestigial remains of a pre-Christian ritual enacting the death and resurrection of a sun god at the time of the Winter Solstice, despite no evidence to support this notion.[6] The plays continue to exist today in many forms, contributing to the winter *repertoires* of Morris dancers' sides.[7] Like the plays, the Morris dances also continue to be considered vestigial pagan rituals, especially in popular perspectives and publications, despite scholarly refutation.[8] An example of the former can be seen in the New Age writer Michael Bayley's assertion that the characters of the Mummers' Play are all to be found in the night sky along the Milky Way.[9] Bayley's conviction that somehow these are astronomically related to the ancient Celtic sky, whilst unsupported, is typical of the manner in which modern paganism and folkloric customs, however incorrectly associated together, are employed for mutual justification and laying claims to pagan continuity.

4 M. Preston, 'The British Folk Play: An Elaborated Luck Visit', *Western Folklore,* (1971), Vol. 30(1), [hereafter Preston, 'The British Folk Play'], pp. 45-48; Millington, 'Origins and Development', for a full discussion of these early records of the plays.

5 Chambers, *English Folk Play*; Reginald Tiddy, *The Mummers' Play* (Oxford: Clarendon Press, 1923), [hereafter Tiddy *The Mummers' Play*], and Alex Helm, *The English Mummers' Play* (Woodbridge: Folklore Society, 1981). Each give good descriptions of the action and cast from region to region, including information on how the plays persisted and developed into the twentieth century.

6 Chambers, Helm and Tiddy's works on the plays particularly promulgated this perception of paganism, which in turn has led to a quest for origins.

7 An Internet search for the Morris will reveal many such claims and associations.

8 J. Forrest, *The History of Morris Dancing, 1458-1750* (Cambridge: James Clarke, 1999).

9 Michael Bayley, *Caer Sidhe, Volume 1: The Celtic Night Sky* (Chieveley: Capall Bann, 1997), p. 55.

The early association of folklore studies with anthropology, from the time of these disciplines' beginnings in the nineteenth century, appears to have greatly influenced the pagan origin argument for folkloric customs. The anthropological theory of evolutionary survivals of Tylor and Frazer has served to form the template upon which to view other cultures.[10] Cultures were considered to be evolving through stages of magic and superstitious rituals, on to organised religions such as Christianity, and finally to the then popular revelation of scientific verity. It suggested that customs such as the Mummers' Play, with its distinctive themes and motifs, could be recognised in primitive cultures elsewhere in the world, and must therefore have been a relic of these earlier cultural forms in our own society. It was a tenacious and all-embracing theory, yet was eventually overturned and replaced in anthropology by a succession of alternative theories.[11] It has surprisingly persisted amongst folklore scholars until the late twentieth century.

The idea that the Mummers' Play was an evolutionary survival did not stand up to academic scrutiny; the cracks began to appear from the 1980s. Folklorists then began to accept the rejection of the theory and recognised not only how traditions were constantly being (re) invented, but how there was absolutely no evidence for the existence of the Mummers' Play prior to the first records in the mid-eighteenth century.[12] Preston's suggestion that the plays were added to existing luck-bringing, house-visiting customs, and then quickly disseminated around the country via some unknown agency, is upheld by contemporary folk play scholars.[13]

Luck-visits and non-evolutionary survivals

Luck-visiting customs feature in the works of Roslyn Frank.[14] She has systematically studied them as survivals of cognitive linguistic cultural traditions. This is altogether different from the old ideas of evolutionary survivals. For Frank, modern luck-visiting customs can be traced through a persistence of association, cultural metaphor and coded references to an

---

10 Edward Burnett Tylor, *Primitive Culture*, 2 Volumes (London: John Murray, 1871); and Sir James Frazer, *The Golden Bough: A Study in Magic and Religion*, 12 Volumes (London: Macmillan, 1911-15).

11 Wendy James provides a modern overview of these alternatives in *The Ceremonial Animal: A New Portrait of Anthropology* (Oxford: Oxford University Press, 2004).

12 Eric Hobsbawm and Terence Ranger, *The Invention of Tradition* (Cambridge: Cambridge University Press, 1983).

13 Preston, 'The British Folk Play'; and Millington, 'Origins and Development'.

14 Roslyn Frank, 'Hunting the European Sky Bears: When Bears Ruled the Earth and Guarded the Gate of Heaven', in V. Koleva and D. Kolev, eds., *Astronomical Traditions in Past Cultures* (Sofia: Sofia Institute of Astronomy, Bulgarian Academy of Sciences, 1996), pp. 116-42; and Roslyn Frank and Mikel Susperregi, 'Conflicting Identities: A Comparative Study of Non-Commensurate Root Metaphors in Basque and European Image Schemata', *Language and Identity*, (2001), pp. 135-83.

earlier cosmovision, when the bear cults of Europe prevailed. Her work identifies elements of these early traditions as continuations in European luck visits and in particular the British Mummers' Plays. There is, too, an astronomical basis argued for by Frank, in that these plays can be traced back to a reverence toward the Bear constellations around the North Pole. Frank's thinking gives some support to the luck visits having a pre-Christian origin, but does not necessarily imply that the play, added later, can be located in the same cognitive linguistic field.

Frank's work succeeds, however, in putting forward an alternative survival theory. Non-evolutionary survivals certainly exist if metaphoric and symbolic associations are employed when tracing persistent cultural forms. For example, the coronation ceremony is one such form that, whilst part of the British Anglican Church ceremonial, can be shown to have a pagan origin.[15] The placing of the golden crown upon the monarch's head originally denoted divine rule and symbolically represented the sun god *Sol Invictus*. The associations take us back to the rule of the pagan Roman emperor Aurelian.[16]

Looking at survivals thus, the symbolic and associational meaning is retained through cultural forms and in particular through collective imagery and linguistic presence. This is not unlike Bakhtin's literary concept of retained cultural relics in carnival and other customs, or Seznec's examination of pagan continuity through literature and art, or what Husserl had described as a sediment of earlier forms.[17]

A new reading-intertextuality

The present position in thinking about the plays leaves many questions unanswered. There is no current explanation as to why a play, with elements seemingly from the Crusades,[18] has the a-historical and incongruous addition of a quack doctor and then a ceremonial collection conducted by an assortment of characters who seem to be echoes from earlier morality plays;[19] or why, also, there is a death and resurrection theme coinciding with the Christmas (or Solstice) period. These questions focus on two approaches: either these components are explained by a steady, accidental accumulation of *strata* from previous cultural forms, or

---

15 F.W. Maitland, *The Constitutional History of the English* (Cambridge: Cambridge University Press, 2008).

16 W. Smith, *A Dictionary of Greek and Roman Antiquities* (London: John Murray, 1875).

17 M. Bakhtin, *Rabelais and His World*, trans. Helene Iswolsky (Bloomington: Indiana University Press, 1984); J. Seznec, *The Survival of the Pagan Gods*, Bollingen Series XXXVIII, trans. B.F. Sessions (New York: Princeton University Press, 1972[1953]); For Husserl's thinking on sediment see Edmund Husserl, *Origin of Geometry*, trans. John Leavy (London and Lincoln: University of Nebraska Press/Bison, 1989).

18 See particularly Tiddy, *The Mummers' Play;* and Chambers, *English Folk Play.*

19 A. Pollard, *English Miracle Plays, Moralities, and Interludes* (Oxford: Clarendon Press, 1946 [1890]).

they were deliberately put together for some purpose and then added to the visiting customs in the eighteenth century. The historical records suggest a deliberation.

In order to probe further, some consideration of a methodology is required, and one which is capable of penetrating through the layers of these intertextual strata. The folklorists and folk play scholars Eddie Cass and Steve Roud conceded that 'the scripts and words themselves must surely hold the answers, if only we knew how to read them'.[20] The method I adopted encompassed a latent textual analysis looking for coded meaning through close reading, as opposed to previous approaches. For example, Millington, by using a manifest textual analysis, missed the interpretation that sending St George to Jamaica was no different to the other variant destinations such as cook shop, pastry cook and kitchens.[21] The close reading recognises the latent meaning: 'Jam-maker' (Jamaica) as a pun on existing variants, and by so doing is better placed to recognise hidden meaning (in this case, the pun).

Further, analysis required an in-depth knowledge of the processes of symbolic associations as expounded in the psychoanalytic works of Freud and Lacan.[22] The function of the unconscious in the creation of texts and their inevitable relationship with other texts (more readily known as intertextuality), became increasingly significant.[23] These elements of interpretative reading came together in the Structuralist-Hermeneutics of the French literary theorist Michael Riffaterre, whose method demonstrates the presence of hidden meaning in a literary work through association and symbol within the text. It works in the same manner as the psychoanalytic interpretation of the symptom can be traced associationally back to its repressed meaning.[24]

20 E. Cass and S. Roud, *Room, Room, Ladies and Gentlemen: An Introduction to the English Mummers' Play* (London: English Folk Song and Dance Society, 2002), p. 18.

21 Millington, 'Origins and Development', used a manifest textual analysis of the play scripts coding line variants. In the lines where the Turkish Knight says he will 'hash thee and smash thee as small as flies and send thee to the "cook shop" to make mince pies', there are many variants, such as 'pastry cook', 'kitchens' and 'Jamaica'. Millington coded the kitchen themes together, but argued that Jamaica was entirely different.

22 Sigmund Freud, *The Interpretation of Dreams* (Ware: Wordsworth Editions, 1997 [1900]); Lacan is perhaps foremost in this type of analysis as he moved on from Freud's drive theories to consider the whole of the unconscious processes linguistically and functioning like a language. For Lacan, symbolic grammar operates through metaphor and metonymy, see Jacques Lacan, *Ecrits: A Selection* (London: Routledge, 1977).

23 For intertextuality see Julia Kristeva's *Revolution in Poetic Language* (New York: Columbia University Press, 1984), [hereafter Kristeva, *Revolution in Poetic Language*]; and Graham Allen, *Intertextuality* (London: Routledge, 2000).

24 M. Riffaterre, *Semiotics of Poetry* (Bloomington: Indiana University Press, 1978), and also M. Riffaterre, *Text Production* (New York: Columbia University Press, 1983).

Earlier I mentioned the problem of whether the plays' genesis and construction was deliberate or accidental. This offers a major dichotomy of direction throughout this study. Thus far, interpretation is considered possible from an authorial intention where meaning is coded deliberately. Literary theorists also acknowledge that, in the field of intertextuality, texts make links with other texts, not just through the author's unconscious, but they can do so through their cultural literary inheritance — texts might seem to write themselves, as Kristeva and Barthes and Derrida in particular have asserted.[25] This could mean that if there is any hidden meaning in the plays' texts veiling the why, how and who of their creation, such presences might not necessarily be a deliberate device, but could appear through these more unconscious inter-textual processes present in their writing.

The oft-named father of linguistics, Ferdinand de Saussure, in fact spent considerable time trying to sort out this dichotomous problem. Saussure was familiar with a phenomenon evident in classical poetry, in which anagrams of the poem's theme (a goddess or god, terrestrial or astronomical feature, for example) are to be found woven throughout the work.[26] After several years trying to identify the literary rules that these poets might have used — i.e., rules akin to those employed by the cryptic crossword-setter, and the unconscious mechanisms of dream and symptom symbolism — all to no avail, and he was forced to give up. Saussure's main discovery was that in the process of writing, anagrammatical elements of the theme being written about will automatically appear embedded in the text in quite surprising ways. (Most of us will have experienced something of this when, for example, writing an article, elements and phonemes of the underlying topic will appear before us in the choice of adjectives, verbs and nouns we think we are deliberately choosing. Yet, some of that 'choice' is being selected unconsciously with reference to the topic or the 'punch-line' that we are working up to, and will comprise of phonemes or rhymes or other associations relating to that underlying theme).

Returning to the Mummers' Plays' scripts, there could be similar textual clues and resonances hiding in them, indicating significance relating to their origin and purpose. These might not necessarily have been put there deliberately by the eighteenth century play recorders or authors — they could have manifested unconsciously. They could also draw upon associated intertexts, which in turn could include overt or embedded sources of many

25 This is a post-structuralist viewpoint as promulgated in Kristeva, *Revolution in Poetic Language*; Roland Barthes, 'The Death of the Author', in *Image, Music, Text.* trans. Stephen Heath (London: Fontana, 1977); and Jacques Derrida, *Writing and Difference,* trans. Alan Bass (Chicago: Chicago University Press, 1967).

26 The best study on Saussure's anagram work is Jean Starobinski's *Words on Words: The Anagrams of Ferdinand de Saussure*, trans. Olivia Emmett (New Haven: Yale University Press, 1979).

cultural and linguistic forms, including pagan ones. An example of the kind of work I am doing using this approach follows.

Troy — a signification lock

The three principal characters in the Mummers' Play are St George, Alexander the Turkish Knight and the Doctor. Apart from possible crusade significance, George and Alexander might be connected, but there is nothing in any George myth or tale in which the presence of a doctor is found. There is therefore an a-historical mix of characters with no apparent underlying logical connection. However, in the intertextual field there is such a connection: there are lines in most of the plays where St George speaks of Sabra, the King of Egypt's daughter, a theme which does not appear in early tales of St George, but does so in Johnson's sixteenth century romantic retelling.[27] In Johnson, clearly influenced by the Trojan origin myth of the British as related by Geoffrey of Monmouth, St George is born in Coventry sometime shortly after the founding of England by Brutus, a refugee from Troy.[28] Johnson makes George the son of the High Steward of England, and the King's own daughter, thus he is descended through his mother from Aeneas.

There is no Turkish Knight in Johnson, although there is the Morocco King, yet the action focuses more on killing a dragon, whereas in the folk play there is no dragon, but George fights with Alexander. Who is Alexander? In pursuing this by associational methods, let us consider Paris, the Trojan and an exemplary Turkish Knight, whose alternative name was also Alexander. The myth tells us that a bear suckled the young Paris-Alexander. The significance here is that both St George and Alexander the Turkish Knight have intertextual links with Troy and the legendary origin of the British.[29]

The Doctor poses more problems — the quack is very much a product of the seventeenth and eighteenth centuries,[30] yet the early plays have him using the herb Elecampane as the restorative and mysterious potion for bringing the dead player back to life.[31] Once again there is a Trojan link, for

27 Richard Johnson's *The Seven Champions of Christendom* (1596), is recently re-published: J. Fellows, ed., *The Seven Champions of Christendom, 1596-7* (Ashgate Publications, 2003).

28 A useful version of the Galfridian work is L. Thorpe, trans., *The History of the Kings of Britain: Geoffrey of Monmouth* (Harmondsworth: Penguin, 1973).

29 See entries for Paris-Alexander in William Smith, *Classical Dictionary* (London: John Murray, 1864).

30 Roy Porter's *Quacks, Fakers, and Charlatans in Medicine* (Stroud: Tempus, 2003), illustrates this nicely.

31 The earliest recorded plays are conveniently reproduced by Peter Millington and the Sheffield Traditional Drama Group, accessible on the *Folk Play Research Home Page* (Sheffield University, 2005), http://www.folkplay.info/ [accessed 20 July 2009], [hereafter the *Folk Play Research Home Page*].

Elecampane is the herb plucked by Helen of Troy when she was abducted by Paris-Alexander, hence its Latin name, *Innula Helenium.*[32]

Trojan British origins, riddles and Druidic sympathisers

The three principal characters in the Mummers' Play are thus connected through this obtuse intertextual relationship with Troy, the legendary origin of the British according to Geoffrey of Monmouth, and of particular significance to eighteenth century Celticists and Druid sympathisers, such as Sir William Stukeley, Iolo Morgannwg and others.[33] I refer to such a hidden connection as a 'signification lock'. In this case it suggests that those devising the play for enhancing the Mummers' house visits were using a characterisation disguising something important to them, and something that related to an eighteenth-century perception of British, Trojan myth, or perhaps Druidism as it was perceived and constructed at that time. There is a hint too of an integral recognition of the luck-visit's ancestral bear that Frank has suggested. Further, the whole *façade* of the plays is likely to be a riddle whereby, having decoded thus far, the next stage is to identify which Celtic myth or tale provided the plot for the hero-combat and death and resurrection of the play.

Scholars of Druidic and Bardic lore, and what Hutton calls 'Druidry', have argued whether there was any esoteric lore cryptographically embedded in the literary heritage of the bards.[34] Whether there is or not,

---

32 According to John Gerard, *Gerard's Herbal: John Gerard's Historie of Plants* (Senate Books, 1994[1597]); although other sources, e.g., Florence Ranson, *British Herbs* (Harmondsworth: Penguin, 1949), p. 56, say that the plant grew from Helen's tears when Paris took her.

33 The literary tradition of the Trojan origin of the British is also found in Druidic lore in Alwyn and Brinley Rees, *Celtic Heritage* (London: Thames and Hudson, 1989), and is documented in Lewis Morris, *Celtic Remains* (London: Parker, 1878), [hereafter Morris, *Celtic Remains*], and discussed in Ronald Hutton's *The Druids* (London: Hambledon Continuum, 2007).

34 Whilst the presence of riddles in Druidic literature is fairly well accepted — see Rachael Bromwich, *Trioedd Ynys Prydein: The Triads of the Island of Britain* (Cardiff: University of Wales, 2006), [hereafter Bromwich, *Trioedd Ynys Prydein*], and for embedded meanings see John Minahane, *The Christian Druids: on the Filid or Philosopher Poets of Ireland* (Dublin: Sana Press, 1993) — whether there is an esoteric Druidical system comparable to the Qabalah is contentious. Such has been proposed by William Owen-Pughe, *The Cambrian Biography or Historical Notices of Celebrated Men Among the Ancient Britons* (London: E. Williams, 1803), and Iolo Morgannwg, trans., William Ab Ithel, *Barddas, or a Collection of Original Documents Illustrative of the Theology, Wisdom, and Usages of the Bardo-Druidic System of the Isles of Britain* (Llandovery: Society for Publishing Ancient Welsh Manuscripts, 1862), and also Matthew Arnold, *The Study of Celtic Literature* (New York: E.P. Dutton & Co, 1910). The same has been strongly refuted by D.W. Nash *Taliesin and the Bards and Druids of Britain* (London: J.R. Smith, 1858), [hereafter Nash, *Taliesin and the Bards and Druids of Britain*], and A.L. Owen, *The Famous Druids* (Oxford: Oxford University Press, 1962),

some eighteenth century British Druidic sympathisers were convinced of its existence and probably invented it. Riddles were certainly deliberately included in the authentic Bardic repertoires, as can be seen in *Hanes Taliesin,* translated by many scholars, including Lady Charlotte Guest and Patrick Ford.[35] Nash and Owen were not convinced,[36] but the twentieth century poet Robert Graves later attempted to decode many of these riddles.[37] For example, in *Hanes Taliesin,* the line: 'I was in the Ark with Noah and Alpha'. Graves' torturous explanation suggested the figure Hu Gadarn. However, given the extent of the Bards' knowledge of Greek, Latin and Hebrew, as well as their own tongues, this particular riddle is easily solved, taking into consideration my methodological points on Saussure's anagrams and cryptic crossword clues.[38] We must ask who was with Noah and Alpha in the Ark? The 'I', the 'N' and the 'OA' of Noah, and the additional 'A' of Alpha, speak to me of the re-ordered name *IONA,* the Hebrew for 'Dove', which seems much more likely to be present in the Ark, according to biblical myth, than Hu Gadarn. This method may be applied to the other riddles. If understood in the eighteenth century, these processes could be mimicked in developing a bogus esoteric system.

Celtic myth as a pattern for the play

If eighteenth century Druid sympathisers were devising a disguised Hero-Combat play for attaching to a vestigial pagan luck-visit custom, which of the many episodes of hero-combat in Celtic myth and literature had they chosen?

Following a systematic survey of criteria, the one story, or at least its final section that also has the Doctor and resurrection sequence, is that of *Math Vab Mathonwy*, the Fourth Branch of the *Mabinogion* suite of tales.[39] Here the 'hero', Lleu-Llaw-Gyffes, has a battle with Goronwy (who took Lleu's woman), and is slain in a ritualised manner. His death can only occur if he happens to be standing with one foot on the back of a buck-goat, the other on the rim of a bath of water, and killed by a spear made only when

---

[hereafter Owen, *The Famous Druids*].

35 Patrick Ford, *Ystoria Taliesin* (Cardiff: University of Wales Press, 1992), [hereafter Ford, *Ystoria Taliesin*] is a helpful translation. Lady Charlotte Guest's is found in *The Mabinogion from the Llyfr Coch o Hergest, and Other Ancient Welsh Manuscripts* (Longacre: Ballantyne Press, 1902[1838]).

36 Nash, *Taliesin and the Bards and Druids of Britain;* and Owen, *The Famous Druids.*

37 See Robert Graves' *The White Goddess* (London: Faber, 1977[1961]), in which he argues for an esoteric Bardic system and attempts to decode many of these riddles, the Noah and Alpha being found on pp. 87-92.

38 Stuart Piggot, *The Druids* (Harmondsworth: Pelican, 1968), pp. 36, 106.

39 See any translation of the *Mabinogion;* Ford, *Ystoria Taliesin,* is useful, and also Jeffrey Gantz, *Mabinogion* (Harmondsworth: Penguin, 1976), [hereafter, Gantz, *Mabinogion*].

others are at Church and taking a year to make.[40] All these criteria are fulfilled in the story and Goronwy then kills Lleu.

It is possible to make an astronomical interpretation that this *geas* on Lleu veils a winter solstice or other calendrical allegory: an annual spear (made by non-churchgoers, pagans) could be read as a sunbeam or shaft of sunlight on a specific occasion – in this case when the sun hits the cusp of the Goat (constellation Capricorn) and the Bath (constellation Aquarius).

Following Lleu's death, Gwyddion the magician, who is Lleu's uncle (or possibly foster-father), with the aid of some physicians, manages to bring him back to life and a year later Lleu then slays Goronwy. The Mummers' Play plot is all there and there is also a bear association for Math, the eponymous overlord. Math is father of Gwyddion and has a Celtic name that can mean 'plenty' but can also mean 'bear'.

Astronomy

A tentative astronomical assignation of Math to the constellation *Ursa Major* demands further substantiation. There are astronomical references given to other characters in this branch of the *Mabinogion*: Arianrhod, Gwyddion's sister, is identified with *Corona Borealis*; Gwyddion, in the *Triads*, is assigned to the Milky Way;[41] and Lleu has been assigned to *Perseus*. But this is no indication for any embedded coherent astronomical function of the story, and these assignations could be quite arbitrary.[42]

Similarly, the suggested solstitial symbolism of Lleu's death, *geas*, when calculating precessional considerations, would have dated the original event to approximately 4,000 to 2000 BCE,[43] a somewhat problematical proposal, considering the *Mabinogion*, although based on earlier oral traditions, was not written until at least the thirteenth century CE.[44]

There is perhaps a literary precedent for such a long-remembered event appearing in the same century of the second millenium CE, which John Darrah has pointed out. Geoffrey of Monmouth wrote about a myth that had retained a cultural (and intertextual) survival when he referred to the Giants Dance (now recognised as Stonehenge), and the fetching of its stones

---

40 Dr Mark Williams of Peterhouse, Cambridge, has since advised me that an earlier interpretation of the word used for buck goat in the *Llyfr Coch o Hergest* refers to a white horse, not a goat, which would nullify any claim for pre-
eighteenth century astronomical symbolism being embedded. However, by the eighteenth century it was being interpreted as a goat, and the astronomical symbolism could have been read as such by the persons desiring to disguise and fabricate a play to enhance the luck-visits.

41 See Bromwich, *Trioedd Ynys Prydein.*

42 Martin Griffiths, *Under a Celtic Sky, the Lesser-Known Stories of the Stars*, Lab-Lit article 341 (University of Glamorgan, 2008).

43 André Berger, 'Obliquity and Precession for the Last Five Million Years', *Astronomy and Astrophysics*, (1976), Vol. 51, p. 127.

44 Gantz, *Mabinogion*, p. 51.

from Ireland. The bluestones have long since been identified as coming from Prescelly in Wales,[45] an area that Gruffydd has confirmed was formerly populated with Irish-speaking Celts, so these stones have come from an 'Ireland'.[46] Darvill and Wainwright have succesfully re-dated the erection of these stones to 2300 BCE, a date that is cognate with the suggested solstice configuration. Thus the point that Darrah made regarding such persistent cultural themic memories could also be applied to the suggested solstice symbolism in *Math.*[47]

It is, however, more likely that such readings of *Math,* rather than authorially intended, may be the product of hermeneutic approximations and accidents of intertextuality. Yet such modes of interpretation would be particularly familiar to eighteenth century Bardic enthusiasts.

Lewis Morris — a possibility

There is one small clue as to who such an enthusiast might have been. It would have to be someone with a deep knowledge of the Bardic literature, a belief in the Trojan origins of the British and a motive for devising a play with an English Christian Saint disguising a strong Celtic tradition. Eighteenth century persons with those qualities tended to belong to Welsh societies such as the Ancient Britons and the *Cymmrodorions* and reconstructed Druid orders.

Amongst possible candidates, the figure of Lewis Morris surfaces — he holds all those required attributes.[48] On studying his many papers and poems, a particularly pointed biblical-styled parody was found, in which Morris is bitterly complaining of unfair treatment over his mining operations in Ceredigion.[49] It seems that friends of the Mackworth mining ventures had attacked and robbed Morris, and the lines narrate this story, with Morris and the others given parodic old testament-style names and locations.[50]

45 Frank Stevens, *Stonehenge Today and Yesterday* (London: HMSO, 1936); and the most recent dating by Darville and Wainwright (unpublished, 2008).

46 William John Gruffydd, *Math Vab Mathonwy, An Inquiry into the Origins and Development of the Fourth Branch of the Mabinogi, with the Text and a Translation* (Cardiff: University of Wales Press, 1928).

47 John Darrah, *The Real Camelot: Paganism and Arthurian Romances* (London: Thames and Hudson, 1981).

48 See Morris, *Celtic Remains*; for general information on the Cymmrodorion society, see E. Jones and D. Watkin Powell, *The Honourable Society of Cymmrodorion: A Concise History, 1751-2001* (Aberystwyth: Cambrian Publications/Argraffwyr Cambrian, 2001).

49 Located in a selection of important works and biography on Morris by Hugh Owen, *The Life and Works of Lewis Morris (Llewelyn Ddu o Fôn), 1701-1765* (Anglesey: Anglesey Antiquarian Society and Field Club, 1951), [hereafter Owen, *The Life and Works of Lewis Morris*].

50 The particular piece from Morris is on p. 53 of Owen, *The Life and Works of Lewis Morris*, and is titled 'The 1st book of the Cronicles of ye mines', listed as Folio 83, BM

Of particular significance is the striking similarity in the following lines of Morris' version with an element of the early Whitehaven and Newcastle versions of the Mummers' Play.

Morris:

> Thy face shines like a polished freestone, and thy forehead like cast brass, thy teeth are like a forest of old oak and thy nose like the mountain Vesuvius.

Whitehaven (& Newcastle) earliest play scripts:

> Sir, to express thy beauty, I am no able,
> For thy face shines like the very kitchen table;
> Thy teeth are no whiter than the charcoal,
> And thy breath stinks like the devil's a-se h-le.[51]

Morris' connections with miners, who at that period were a mobile workforce circulating between collieries and mineral mines throughout the north, indicates some mutual influence here. In addition, Morris was also noted for his naval ventures and customs-work that occasionally took him to Whitehaven from Anglesey. Morris would have been familiar with miners' friendly societies, which played host to the Sword Dance versions of the plays.[52]

Given his animosity for Mackworth (who incidentally was a founder member of the Society for the Preservation of Christian Knowledge), a potential motive and scenario for a pagan-themed play, disguised superficially as Christian, and attached to luck-visits (by miners' societies for their funds), begins to emerge.

ADDL. MS.14929, held at Aberystwyth, National Library of Wales.

51 These scripts are reproduced for convenient research purposes by the *Folk Play Research Home Page*. The Newcastle one is nominally the earlier, dated 1746-1769, but there is textual evidence suggesting that the Whitehaven one recorded at 1810-1826 could be the earlier, in Michael Preston, et al., *Chapbooks and Traditional Drama: An Examination of Chapbooks Containing Traditional Play Texts, Part 1: Alexander and the King of Egypt Chapbooks* (CECTAL Bibliographical and Special Series 2, 1977).

52 This is attested in original supporting notes to some Sword Dance Play scripts available on the *Folk Play Research Home Page*, and also dealt with in some depth by Violet Alford, *Sword Dance and Drama* (London: Merlin Press, 1962). Alford's central thesis about links with ancient mining is generally discredited, but the association with miners' groups has significance in the light of this present work. Richard Wolfram's 'Sword Dances and Secret Societies', *Journal of the English Folk Dance and Song Society,* (1932), Vol. 1, pp. 34-41, looks at the pitmen's secret societies.

Conclusion

The mystery of the origin and purpose of the Mummers' Play has been hampered by successions of scholars' and lay persons' insistence for perceiving paganism behind its cultural form. This paper has begun to demonstrate that the Mummers' Play can be read to reveal such a pagan 'fingerprint', but not through direct survival or deliberation. It is one that is perceived through an intertextual and hermeneutic interpretation of sedimentary cultural presence put together in the eighteenth century, most likely by Celtic sympathisers and reconstructionists.

Astronomical interpretations, popular with those seeking pagan continuities, might also be tenuously discerned, yet not robustly enough to assert a pre-Christian continuity. Such astronomical interpretations are cognate, however, with the Celtic lore and knowledge that was at the disposal of eighteenth century Celtic reconstructionists.

The riddle of the Trojan (British) origins of the principal players, once recognised as such, led to a search for a British, possibly Druidic, myth or story as prototype for the plot of the play. The selected candidate, *Math Vab Mathonwy,* holds a veiled presence of the luck-visit bear identified in Roslyn Frank's studies. This story offers therefore an intertextual link uniting the Mummers' pre-existing luck-visits with the play that was added to them in the eighteenth century.

The focus on Celtic personalities led to the figure of Lewis Morris and his possible role in the genesis and development of the Mummers' Play, something that would not have so readily emerged without recourse to the structuralist-hermeneutic approach adopted for decoding the intertextual *strata* of the texts.

Bibliography

Alford, V., *Sword Dance and Drama* (London: Merlin Press, 1962).

Arnold, M., *The Study of Celtic Literature* (New York: E.P. Dutton & Co, 1910).

Bayley, M., *Caer Sidhe, Volume 1: The Celtic Night Sky* (Chieveley: Capall Bann, 1997).

Bakhtin, M., *Rabelais and his World,* Helene Iswolsky, trans. (Bloomington: Indiana University Press, 1984).

Barthes, R., 'The Death of the Author', *Image, Music, Text,* Stephen Heath, trans. (London: Fontana, 1977).

Berger, A.L., 'Obliquity and Precession for the Last Five Million Years', *Astronmy and Astrophysics,* (1976), pp. 51-127.

Bromwich, R., *Trioedd Ynys Prydein: The Triads of the Island of Britain,* Third edition, Rachel Bromwich, ed. and trans. (Cardiff: University of Wales Press, 2006).

Cass, E. and S. Roud, *Room, Room, Ladies and Gentlemen: An Introduction to the English Mummers' Play* (London: EFSD, 2002).

Buckland, T. and J. Wood, eds., *Aspects of British Calendar Customs* (Sheffield: Sheffield Academic Press, 1993).

Chambers, E.K., *The English Folk Play* (Oxford: Oxford University Press, 1933).

Darrah, J., *The Real Camelot: Paganism and Arthurian Romances* (London: Thames and Hudson, 1981).

Forrest, J., *The History of Morris Dancing, 1458-1750* (Cambridge: James Clarke, 1999).

Ford, P.K., *Ystoria Taliesin* (Cardiff: University of Wales Press, 1992).

Frank, R.M., and M. Susperregi, 'Conflicting Identities: A Comparative Study of Non-Commensurate Root Metaphors in Basque and European Image Schemata', *Language and Identity*, (2001), pp. 135-83.

Frank, R. 'Hunting the European Sky Bears: When Bears Ruled the Earth and Guarded the Gate of Heaven', in V. Koleva and D. Kolev, eds., *Astronomical Traditions in Past Cultures* (Sofia: Sofia Institute of Astronomy, Bulgarian Academy of Sciences,1996), pp. 116-42.

Frazer, J.G., *The Golden Bough: A Study in Magic and Religion*, 12 Volumes (London: Macmillan, 1911-1915).

Freud, S., *The Interpretation of Dreams* (Ware: Wordsworth Editions, 1997[1900]).

Gantz, J., trans., *Mabinogion* (Harmondsworth: Penguin, 1976).

Graves, R., *The White Goddess* (London: Faber, 1977[1961]).

Griffiths, M., *Under a Celtic Sky: The Lesser-Known Stories of the Stars.* Lab-Lit article 341 (University of Glamorgan, 2008).

Gruffydd, W.J., *Math Vab Mathonwy: An Inquiry into the Origins and Development of the Fourth Branch of the Mabinogi, With the Text and a Translation* (Cardiff: University of Wales Press, 1928).

Guest, C., *The Mabinogion from the Llyfr Coch o Hergest, and Other Ancient Welsh Manuscripts* (Longacre: Ballantyne Press, 1902[1838]).

Helm, A., *The English Mummers' Play* (Woodbridge: Folklore Society, 1981).

Hobsbawm, E., and T. Ranger, *The Invention of Tradition* (Cambridge: Cambridge University Press, 1983).

Husserl, E., *Origin of Geometry*, J.P. Leavy, trans. (London and Lincoln: University of Nebraska Press/Bison, 1989).

Hutton, R., *The Stations of the Sun: A History of the Ritual Year in Britain* (Oxford: Oxford University Press, 1996).

Hutton, R., *The Origins of Modern Druidry: Mount Haemus Lecture for the Year 2005* (Lewes: The Order of Bards, Ovates and Druids, 2005).

Hutton, R., *The Druids* (London: Hambledon Continuum, 2007).

Gerard, J., *Gerard's Herbal: John Gerard's Historie of Plants* (Senate Books, 1994[1597]).

James, W., *The Ceremonial Animal: A New Portrait of Anthropology* (Oxford: Oxford University Press, 2004).

Johnson, R., *The Seven Champions of Christendom:* 1596

Johnson, R. and J. Fellows, eds., *The Seven Champions of Christendom, 1596-7* (Ashgate Publications, 2003).

Jones, E., and D. Watkin-Powell, *The Honourable Society of Cymmrodorion: A Concise History, 1751-2001* (Aberystwyth: Cambrian Publications/Argraffwyr Cambrian, 2001).

Kristeva, J., *Revolution in Poetic Language* (New York: Columbia University Press, 1984).

Lacan, J., *Ecrits: A Selection* (London: Routledge, 1977).

Maitland, F.W., *The Constitutional History of the English* (Cambridge: Cambridge University Press, 2007).
Morris, L., *Celtic Remains* (London: Parker, 1878).
Millington, P.T., 'The Origins and Development of English Folk Plays' (PhD Thesis, Sheffield University, 2002).
Minahane, J., *The Christian Druids: on the Filid or Philosopher Poets of Ireland* (Dublin: Sana Press, 1993).
Monmouth, G., *The History of the Kings of Britain: Geoffrey of Monmouth,* L. Thorpe, trans. (Harmondsworth: Penguin, 1973).
Nash, D.W., *Taliesin and the Bards and Druids of Britain* (London: J.R. Smith, 1858).
Owen. A.L., *The Famous Druids* (Oxford: Oxford University Press, 1962).
Owen, H., *The Life and Works of Lewis Morris (Llewelyn Ddu o Fon), 1701-1765* (Anglesey: Anglesey Antiquarian Society and Field Club, 1951).
Owen-Pughe, W., *The Cambrian Biography or Historical Notices of Celebrated Men Among the Ancient Britons* (London: E. Williams, 1803).
Piggot, S., *The Druids* (Harmondsworth: Pelican, 1968).
Pollard, A.W., *English Miracle Plays, Moralities, and Interludes* (Oxford: Clarendon Press, 1946[1890]).
Porter, R., *Quacks, Fakers and Charlatans in Medicine* (Stroud: Tempus, 2003).
Preston, M.J., 'The British Folk Play: An Elaborated Luck Visit', *Western Folklore,* (1971), Vol. 30(1), pp. 45-48.
Preston, M.J., M.G. Smith and P.S. Smith, *Chapbooks and Traditional Drama: An Examination of Chapbooks Containing Traditional Play Texts, Part 1: Alexander and the King of Egypt* (CECTAL Bibliographical and Special Series 2, 1977).
Rees, A. and B. Rees, *Celtic Heritage* (London: Thames and Hudson, 1989).
Ranson, F., *British Herbs* (Harmondsworth: Penguin, 1949).
Riffaterre, M., *Semiotics of Poetry* (Bloomington: Indiana University Press, 1978).
Riffaterre, M., *Text Production* (New York: Columbia University Press, 1983).
Seznec, J., *The Survival of the Pagan Gods*, Bollingen Series XXXVIII, B.F. Sessions, trans. (New York: Princeton University Press, 1972[1953]).
Simpson, J. and S. Roud, *Dictionary of English Folklore* (Oxford: Oxford University Press, 2003).
Smith, W., *Classical Dictionary* (London: John Murray, 1864).
Smith, W., *A Dictionary of Greek and Roman Antiquities* (London: John Murray, 1875).
Starobinski, J., *Words on Words: the Anagrams of Ferdinand De Saussur*, O. Emmet, trans. (New Haven: Yale University Press, 1979).
Stevens, F., *Stonehenge Today and Yesterday* (London: HMSO, 1936).
Tiddy, R.V.E., *The Mummers' Play* (Oxford: Clarendon Press, 1923).
Traditional Drama Research Group, on the *Folk Play Research Home Page* (Sheffield University, 2005), http://www.folkplay.info/ [accessed 20 July 2009].
Tylor, E.B., *Primitive Culture*, 2 Volumes (London: John Murray, 1871).
William Ab Ithel, trans., *Barddas (Iolo Morgannwg) or a Collection of Original Documents Illistrative of the Theology, Wisdom, and Usages of the Bardo-Druidic System of the Isles of Britain* (Llandovery: Society for Publishing Ancient Welsh Manuscripts, 1862).
Wolfram, R., 'Sword Dances and Secret Societies', *Journal of the English Folk Dance and Song Society,* (1932), Vol. 1, pp. 34-41.

The Beltane Fire Festival: its Place in a Contemporary World

Pauline Bambrey

It is widely believed that the four fire or quarter-day festivals — Samhuinn, Imbolc, Beltane and Lughnasadh — were of great importance and significance in pre-Christian communities. The festivals brought communities together to strengthen bonds, re-establish identities and celebrate the changing seasons. Do these festivals have a role in a contemporary world and, if so, what? This paper will explore these issues by reporting on ethnographic evidence obtained through anthropological research with the Beltane Fire Society based in Edinburgh.

There are many festivals held across the United Kingdom each year and I suggest their organisers and participants would all argue that they have a place within a contemporary world. This paper will consider the place of fire festivals and in particular the festival of Beltane held on the evening of 30 April each year in the city of Edinburgh.[1]

One of the largest celebrations of the festival of Beltane in the United Kingdom is organised by the Beltane Fire Society (BFS) of Edinburgh. Beltane is considered to be a pre-Christian festival celebrated on the eve of 30 April and the day of 1 May.[2] It marks the end of winter and the beginning of summer and is one of the two principal festivals of the year, the other being Samhuinn on 31 October.[3] Beltane is a time for rejoicing and celebrating the survival of the winter months,[4] a time for renewal, regeneration, resurrection or rebirth,[5] a festival of life.[6]

1 This is not the only Beltane festival to be held in the UK.

2 G. Webster, *The British Celts and their God under Rome* (London: B.T. Batsford Ltd., 1986), [hereafter Webster, *The British Celts*], p. 32; R. Hutton, *The Pagan Religions of the Ancient British Isles: Their Nature and Legacy* (Oxford: Blackwell, 1991), [hereafter Hutton, *The Pagan Religions of the Ancient British Isles:*], p. 176; E.O. James, *Seasonal Feasts and Festivals* (London: Thames & Hudson, 1961), [hereafter James, *Seasonal Feasts and Festivals*], p. 312; J.A. MacCulloch, *The Religion of the Ancient Celts* (London: Constable, 1911), [hereafter, MacCulloch, *The Religion of the Ancient Celts*], p. 257.

3 MacCulloch, *The Religion of the Ancient Celts*, p. 257; T.D. Kendrick, *The Druids*, p. 130; Hutton, *The Pagan Religions of the Ancient British Isles:*, p. 176; F. Marion McNeill, *The Silver Bough, Volume 2: A Calendar of Scottish National Festivals: Candlemas to Harvest Home* (Glasgow: William MacLellan, 1959), [hereafter McNeill, *The Silver Bough, Volume 2*], p. 56.

4 A. Ross, *Pagan Celtic Britain: Studies in Iconography & Traditions* (London: Routledge

The Beltane fire, or *neid*-fire,[7] was of great importance to the celebratory rituals and were traditionally lit on hilltops.[8] Their use, according to McNeill, 'was twofold – propitiation [sacrifice] and a purification'.[9] Although there is evidence of human sacrifice in the first century BCE, there is no evidence of this happening in Britain. However, McNeill does suggest that there remain traces of sacrifice in the way the surviving ritual is performed, particularly in the 'offering of the cake'.[10] James notes the sacrificial element of the fire and relates the tale of the burning of a horse's skull or bones, which he suggests is a remnant of the sacrifice of a horse 'at the ancient Celtic festival'.[11] MacCulloch, meanwhile, makes reference to the slaying of an animal.[12]

The purification rituals involved the use of fire in various ways. McNeill informs us that members of the community would sprinkle each other with ashes from the fire and torches would be lit from it and carried around fields and homesteads in order to ensure the fertility of their crops and protection of their homes for the coming year. She also suggests that the fires 'purified the air', thus ridding it of 'all malign influences', protecting the community from disease, vermin, and natural calamities (for example, lightening and thunder) and from 'wicked enchantments'.[13]

Other rituals of purification, protection and fertility involved herds and flocks being driven either between or through the fire before being put out to summer pasture. Some sources inform us that two fires are lit with a narrow alley between them in order to drive flocks and herds between them as an act of purification and protection before putting them out to summer pasture.[14] There is conflicting evidence as to whether one or two fires were lit, and whether the animals were driven through or between the fires. It is suggested that people also 'leapt' across the fire for various reasons, such as for protection and luck if they were about to undertake long journeys or go hunting. Young women were said to jump across the fire to aid them in finding a good husband, and pregnant women to ensure an easy and safe delivery.

---

and Kegan Paul, 1967), [hereafter Ross, *Pagan Celtic Britain*], p. 162.

5 Webster, *The British Celts*, p. 32.

6 MacCulloch, *The Religion of the Ancient Celts*, p. 257.

7 James, *Seasonal Feasts and Festivals*, p. 312; MacCulloch, *The Religion of the Ancient Celts*, p. 258; McNeill, *The Silver Bough, Volume 2*, p. 56; Webster, *The British Celts*, p. 32.

8 James, *Seasonal Feasts and Festivals*, p. 213.

9 McNeill, *The Silver Bough, Volume 2*, p. 56.

10 McNeill, *The Silver Bough, Volume 2*, p. 56.

11 James, *Seasonal Feasts and Festivals*, p. 313.

12 MacCulloch, *The Religion of the Ancient Celts*.

13 McNeill, *The Silver Bough, Volume 2*, p. 57.

14 See for example, McNeill, *The Silver Bough, Volume 2*, p. 56; Webster, *The British Celts*, p. 32; MacCulloch, *The Religion of the Ancient Celts*, p. 264; R. Hutton, *The Stations of The Sun: A History of the Ritual Year in Britain*, p. 218.

On the eve of Beltane, all household fires were extinguished and the final task of the torch-bearers was to re-light the fires in their hearths from the sacred Beltane fire, these hearth fires would then be kept alight throughout the coming year.[15]

However, it is important to remember that most of this evidence of pre-Christian times is derived from Roman and Christian sources. The festival traditions, which were passed down orally from generation to generation, will have changed in the telling and, although there may be core elements that have survived, what we know and understand them to be now may be considerably different to how they were actually celebrated – a point of significance when we consider the Beltane festival as celebrated by the BFS.

The Beltane Fire Society of Edinburgh was formed by Angus Farquhar in 1987. Farquhar had been a member of a drumming group called Test Dept who were at that time, to use Farquhar's words, 'politically engaged'.[16] The group toured with the South Wales Striking Miners' Choir across the UK, and Farquhar describes this time as being particularly hard. He witnessed riots and violence between the police and the striking miners and their supporters and, whilst he supported the strike, he felt that this type of action was not the way forward and that there must be a better solution.

Test Dept had also toured extensively across other countries, including Italy and Spain, and Farquhar had been struck by the number of public rituals in the form of festivals, carnivals and pageants which allowed people to express themselves through ritual, whether intended as to celebrate the time of year or a religious feast or bring the community together. He states 'it was the idea of culture as resistance',[17] non-violent resistance, that really led Farquhar to think about what his own home city offered in the way of public ritual. He found that:

> instead of one side always being in opposition to the other [i.e., to the government], what I began to see through folk ritual was the idea that you could make your own world, you can celebrate who you are, it doesn't matter if you've got money, it doesn't matter what else is going on, you have your own sense of community, your own sense of place.[18]

It was at this point that Farquhar left Test Dept, returned to Edinburgh and began to question why there were no such festivals/rituals held there, that surely somewhere within folk history there must have been similar rituals. Farquhar turned to the School of Scottish Studies based in Edinburgh and, with the help of Margaret Bennett and Hamish Henderson, began to

15 James, *Seasonal Feasts and Festivals*, p. 312; Matthews, *Source Book*, p. 354; Ross, *Pagan Celtic Britain*, p. 162.

16 A. Farquhar, Talk to Processional Drummers (15 March 2009), [hereafter Farquhar, Talk to Processional Drummers].

17 Farquhar, Talk to Processional Drummers.

18 Farquhar, Talk to Processional Drummers.

research the subject. Relying on the work of F. Marion McNeill, Bennett, Henderson and Farquhar concluded that Beltane fires were lit and rituals performed on Arthur's Seat, a prominent landmark that now resides within the city of Edinburgh. Ronald Hutton, though, has questioned the claim that Beltane took place in pre-Christian times within the city of Edinburgh, on the grounds that it was a 'pastoral' festival.[19]

However, Farquhar has confidence in McNeill as a well known and respected local historian and he, Bennett and Henderson had no reason to doubt her information. In volume four of *The Silver Bough,* McNeill wrote under 'Edinburgh, Beltane Rites' that:

> ARTHUR'S SEAT, a hill over 800 feet, behind the Palace of Holyroodhouse, is one of the traditional sites on which our pre-Christian forebears were accustomed to light their Beltane fires at sunrise on the first day of May, to hail the coming of summer and to encourage by mimetic magic the renewal of the food supply.[20]

Further, in volume two, under the title 'Sites of the festival', she wrote: 'There are plentiful traces of Beltane sites all over Scotland. Among the best known are Arthur's Seat, near Edinburgh'.[21] McNeill's hypothesis, therefore, is that Beltane fires were lit on Arthur's Seat from pre-Christian times until the mid-nineteenth century.

Farquhar decided that he wanted to 'resurrect' the Beltane festival in Edinburgh but knew it would not be possible to hold it on what he believed to be its original site of Arthur's Seat, an area now owned by the Crown. Henderson suggested holding the festival on Calton Hill instead, as it is known as the people's hill and thought to be where witches once lived and also where the faerie boy of Lleith disappears into the hill.

In 1987, the first 'contemporary' Beltane festival took place on Calton Hill. Angus did not want to simply reproduce what is historically known about the pre-Christian festival. He wanted to revive it,

> not as a pastiche of ancient dramatic forms, but as a living embodiment of an ancient and simultaneously modern spirit ... an invocation both local and universal made from the energy and skills of all who took part with the only bounds being the tightly coiled relationship between the female energy of the May Queen and the male energy of the fire and the building of that spiral of power as it marked the bounds of the site.[22]

---

19 Ronald Hutton comment at the Sophia Centre conference, Bath, 6 June 2009.
20 McNeill, *The Silver Bough, Volume 4: The Local Festivals of Scotland* (Glasgow: Stuart Titles Ltd., 1968), p. 78.
21 McNeill, *The Silver Bough, Volume 2*, p. 72.

---

22 A. Farquhar in email to Beltane Fire Society Members (15 February 2007).

Farquhar enlisted the help of choreographer Lyndsey John from St. Lucia, who brought an international influence to the proceedings with the use of Bhuto, whilst the first May Queen, Liz Rankin, a dancer from DV8 (a contemporary dance company in the 1980s and 1990s) conferred a wild but disciplined female energy. In the early years Test Dept provided the drumming, contributing a regular, deep rhythm — Farquhar describes it as 'the engine', and finds that all these types of ritual have a 'driving force',[23] which was the drumming. The first Beltane celebration in 1987 consisted of a May Queen (Liz Rankin), eight drummers (from Test Dept), two red men and one blue man, and attracted an audience of approximately one hundred people.

Farquhar and Rankin continued to be involved with the BFS for ten years, during which time the festival grew, not only in the number of performers, but also in audience size, and by 1992 there were 10,000 people coming to celebrate the event. By 2002, the audience size had grown to approximately 15,000. In 2003, Edinburgh City Council refused to grant the society a license and, for the first time in sixteen years, there was no Beltane on Calton Hill. The hill was cordoned off and guarded by police on the night of 30 April that year. For the society, that year's celebration was a smaller affair, taking place outside Edinburgh on the night of 2 May in front of a small, invited audience.

After renegotiation, the society was allowed back on to Calton Hill in 2004. It was agreed that Beltane would become a ticketed event (until then there had been no charge to attend the festival) and that ticket numbers would be limited to 12,500, with a curfew of 1am. Many of the society's members felt that making the audience pay to attend went against the whole ethos of a 'public' ritual. However, it was agreed that a nominal charge would have to be made in order for the festival continue on that site, and in 2004 Beltane was back on the hill and has remained there, keeping to the guidelines set out by the Edinburgh City Council.

Today, the BFS has over three hundred performers, and over the years the festival has evolved and expanded. Numbers in the original groups of drummers, red men and blue men have grown and new groups have developed. The cast consists of: a May Queen and a Green Man — the Queen's summer consort; the blue men — keepers of the tradition and holders of the wisdom of the court. The blue men ensure the smooth running of Beltane and pass on the traditions to members of the other groups. They protect and escort the procession as it makes its way around the hill; the white women, or white warriors as they are sometimes known — the May Queen's protectors and handmaidens. They are pure and staid, their faces expressionless and their movements rigid and precise — it is here that we see the biggest influence of Bhuto, also the international influence of white face paint and bright red lips as can be found with the

23 Farquhar, Talk to Processional Drummers.

Japanese geisha. The red men arepolar opposites to the white women. They are not representations of the devil as is often thought. They wear red body paint and red loin cloths and nothing else. They live for the day; they are wild and chaotic in behaviour, with their symbolic lewd and licentious acts. Their goal for the night is to seduce the white women in an effort to bring a balance to the coming season.

In more recent years, due to the increase in the number of participants, elemental groups — earth, air, fire and water — have been formed. These groups have the biggest turnover of members with consequent changes in the nature of the ritual and interpretations of the elements — apart from the presentation of an offering to the May Queen, which remains a constant feature.

The other groups are the processional drummers who, as stated earlier, provide the deep, resonant drumming that accompanies the procession around the hill; red beastie drummers whose drumming is wild and frantic, in keeping with the nature of the group they accompany, the red men; fire arch, who are guardians to the gate of the 'otherworld'; No Point, who act as court jesters, keeping those members of the audience who cannot see what is going on in the main procession occupied — they also 'open up' and keep spaces open ready for groups/processions to move into; torch-bearers, who light the procession as it makes its way around the hill; stewards, who create a physical barrier between the procession and the audience. Last but not least we have two relatively new groups: Green Point (in existence for three years now), who are in charge of collecting and sorting litter for recycling on the night; and Photo Point, the society's official photographers who take the amazing photographs that can be found on the society's website.

So, what do all these people do? How does it all fit together? Let us look briefly now at the proceedings for the night before, which begin with the lighting of the neid-fire on top of the acropolis — from this, all the torches are lit for the night. The procession is led by torch bearers and processional drummers who rise up over the acropolis and descend the steps at the front. The torch bearers are followed by the white women who arranged themselves down the steps at the front of the acropolis awaiting the arrival of the May Queen and Green Man. The May Queen appears on the top steps of the acropolis and, with the white women, performs some ritual movements before descending the steps between her handmaidens, closely followed by the Green Man. The procession gathers, with blue men as escorts at the front, the May Queen and Green Man followed by white women, all flanked by processional drummers, torch bearers and stewards. The procession then makes its way down to the fire arch and passes through it — this is not only symbolic of entering the 'other world', but also of passing between two fires, as in pre- Christian times when cattle and flocks were driven through or between the fires. From here the May Queen and her entourage visit and awaken each elemental point, watching their rituals and accepting their offerings. After awakening each element, the procession continues around the hill, encountering a charge from the red men as they go. This assault is dealt with by the white women and the procession continues until they reach an area where a stage has been erected.

By this time the elemental groups and other groups have congregated at the stage to await the Queen and her entourage; once there, every member is presented to and recognised by the May Queen — before that moment, no individual should have looked the May Queen in the eye. Following this, the May Queen, some of her handmaidens and the Green Man take to the stage (the Green Man at this point is completely covered in greenery, no part of his body is visible). The Green Man stands to the side of the stage whilst the May Queen and her handmaidens begin to spin, getting faster and faster. Suddenly, the Green Man, unable to contain himself, touches the May Queen and instantly falls to the ground, dead. The May Queen stops and steps away, the handmaidens converge upon the body of the Green Man and divest him of his greenery, throwing it to the waiting red men. Once stripped, they lift him aloft and turn him three times before lying him back down on the stage. The May Queen approaches and stands over him, she breathes life back into him and draws him back to the present, making him rise and live again. He stands, recovers and then dances, wildly and ecstatically, with the sheer joy of being given a second chance. At the end of his dance, the May Queen takes his hand and draws him to her, he kneels before her and she crowns him with a crown of greenery, takes his hand and presents him to the others as her summer consort. From here, the two leave the stage and make their way to the bonfire which is waiting to be lit. The Queen and her consort light the bonfire with specially made torches that have been lit from the neid-fire.

After this, the procession makes its way to the bower, where refreshments are awaiting the Queen and her entourage. It is here that everyone congregates and celebrates. Handfastings are performed and, finally, the red men succeed in their mission to seduce the white women, they come together to dance and drink, music is played and everyone dances and rejoices. At this point the audience are allowed to mingle with the performers.

The whole occasion is marked by the energy, commitment, joy and a friendship that emanate throughout this process, both by the performers and the audience. The audience get caught up in what is happening, even if they are unsure of what it is all about — I have heard comments from the audience such as 'Oh my god, this is amazing, I have never felt energy like it' and 'I feel so good, how can something like this make you feel so energised?'

This brings us to the the question of the importance and place of such festivals within the contemporary world. Let us address first the importance of the role of the audience.

Some of the people I have spoken to who were part of the audience state that they felt part of the proceedings, they were not just onlookers. They are made to feel part of it, even if a complete stranger to Beltane were to come and watch and say 'I don't understand, what is it all about?' Other members of the audience are often found explaining what is happening to the newcomers, and this is seen as part of what it is all about; the audience is part of the performance, participating and experiencing by acknowledging and commenting on what is taking place.

Bowie suggests that rituals are 'dramatic', and considers that 'Rituals can be seen as performances, which involve both audience and actors'.[24] One could argue that Beltane as 'performed' by the BFS is a performance rather than a ritual; however, as Schechner suggests, performance is dependant upon its content and function as to whether it is ritual or theatre.[25] He argues that if it has 'efficacy' it is ritual, whereas if it is for entertainment it is theatre. He also argues that no performance is purely one or the other, and the boundaries between them are never fixed or static.

The BFS' interpretation of Beltane includes elements of both efficacy and entertainment, and could be therefore said to span both ritual and theatre. Any audience, be it for ritual or performance, is 'involved' or 'participating', in the sense that it provides affirmation for the actors. De Coppet suggests, that rituals are also for those watching, not just for those performing,[26] and Lewis finds that rituals involve a 'sense of special occasion ... [and call] for public attention'.[27]

During interviews I have undertaken with members of the BFS, they have described their early experiences of being members of the audience as being like nothing else they had ever seen or experienced. As audience menbers, they felt part of the proceedings and were driven by the need to find out more and to experience it through participation as a member. One informant told me that she had been to see Beltane numerous times as a child with her family and could not wait until she was old enough to become a member of the society (there is a minimum age requirement of eighteen years old), and Beltane 2008 was to be her first year as a BFS member. Some members informed me that participating in the festival helped them to reconnect with the seasons, something that can easily be overlooked or missed when living in a big city.

The festival provides a reason for people to congregate, it brings a diverse group of people together for a shared purpose. In a discussion about Goth culture, Hodkinson suggests a shared commonality expressed through identity and shared goals and bonds brings groups of people together.[28] For the performers of Beltane, the commonality is their desire to take part in the festival, and the shared goal is to bring to fruition three months of hard

24 F. Bowie, *The Anthropology of Religion* (Oxford, Blackwell, 2000), p. 151.

25 R. Schechner, *Performance Theory* (New York & London, Routledge, 1994), p. 120.

26 D. de Coppet, ed., *Understanding Ritual* (London: Routledge, 1992), p. 8.

27 G. Lewis, *Day of Shining Red: An Essay in Understanding Ritual* (Cambridge: Cambridge University Press, 1980), p. 7.

28 P. Hodkinson, *Goth: Identity, Style and Subculture* (Oxford & New York: Berg, 2002), p. 7.

work to make the festival a success. Friendships and bonds are created with people who share common interests. Being a BFS member involves being part of a collective identity, and members refer to themelves as 'Beltaners'. However, the sense of belonging is not limited to other members of the BFS or to a particular subdivision within the group, such as white women, blue men, red men or elementals – there is also a sense of belonging to the wider community and collective identity.

The fluidity of the relationship between audience and performer at the Beltane celebration and the frequent transformation of members of the former into the latter is reminscent of Howell's argument that it is possible to belong to more than one community, as well as Cohen's suggestion that our individual identities are reinforced through interactions with others, Brunaker's argument that identity is not a 'thing' that people can have or can be but a process that people 'make' and 'do', and Jenkins' view that identity developes as a continuous process.[29] Cohen finds that rituals with their attendant symbolism help to strengthen the individual social identity and sense of belonging, allowing people to experience community. He includes 'calendrical' or agricultural rituals, a category which includes Beltane, describing them as 'boundary marking rituals', suggesting that they bring communities together, re-establishing and maintaining boundaries.[30]

So how does the ritual aspect of the festival connect to community and identity? We can look at the festival aspect, particularly the changing of the seasons, as a 'rite of passage' of which Van Gennep tells us there are three phases – pre-liminal, liminal and post-liminal. [31] There are certainly elements of the pre-liminal phase when members come together and decide how they are going to be involved in the ritual for that year. The longest phase is the liminal, which corresponds to the period spent in preparation for the festival, when social levelling and group bonding takes place. The final phase, the post-liminal phase, is when individuals reintegrate into communities beyond the celebration of Beltane.

However, the BFS community lives on after the festival. Those who live in Edinburgh meet to socialise, and those who do not reside in Edinburgh communicate via the Internet, through the community's discussion group or on social networking sites such as Facebook. Jon, the BFS chairman, finds that it is this common interest in the festival and their shared experience

29 S. Howell, 'Community Beyond Place: Adoptive Families in Norway', In V. Amit, ed., *Realizing Community: Concepts, social relationships & sentiments* (London & New York: Routledge, 2002); A. Cohen, *The Symbolic Construction of Community* (London: Routledge, 1985), [hereafter Cohen, *The Symbolic Construction of Community*]; R. Brubaker, *Ethnicity Without Groups* (Harvard University Press, 2004), pp. 28-63; R. Jenkins, *Social Identity* (London & New York: Routledge, 2004), p. 17.

30 Cohen, *The Symbolic Construction of Community*, p. 53.

31 A. Van Gennep, *The Rite of Passage* (London: Routledge & Kegan Paul, 1960), pp. 152-153.

that inspires individuals to keep in contact throughout the year, and their connection via the Internet becomes an extension of the physical community of Beltane.

When interviewing members of the BFS, I asked them what was the most important feature of their involvement. Their answers varied – for some the ritual aspect was the most important, for others it was purely the performance. However, the majority stated that it was the friendships they had formed and the sense of belonging to a community of like-minded people that was as important or more important than anything else. Some found the BFS also provided the means to safely explore other aspects of their character. They felt that they were not bound by the usual social conventions; , they could be a red man and wear nothing but red body paint and a loin cloth and would be accepted as part of the proceedings. Other participants stated that being part of Beltane and becoming red, blue, white or elemental allowed them to be who they really were. One participant, a blue man for a number of years, stated 'it is not just surface, it is not just what you are presenting, it's about what *you* are'. Another participant, an ex-blue man, claimed 'it's part of who I am'. One claimed 'Red is an expression of something within you; you're not painting on a character'. Another, also a red man, stated 'It is not you ... it is the red in you', and another similiarly suggested that it 'taps into something that is already there'. Many others claimed that the character they became for Beltane was an extension of their identity, something normally kept hidden.

The majority of my informants claimed that they never actually lose that sense of being a character in the ritual, a part of it always remains.

The earlier conflicts with the city council and complaints from local residents have now been resolved and the large audience of residents as well as visitors to the city is testimony to the festival's importance to the wider community.

The evidence suggests the Beltane festival as celebrated by the BFS is an important and accepted part of Edinburgh culture and serves a purpose for residents and visitors to Edinburgh as well as to those who take part in the festival itself. It creates enduring bonds between like-minded people who have shared history, goals and ideals, allowing people to explore their individual and collective identity and reconnect both spiritually and physically to their environment beyond the city. There is therefore a strong argument to be made for the contemporary value and relevance of the ancient calendar festivals in general.

Bibliography

Bell, C., *Ritual: Perspectives and Dimensions* (Oxford: Oxford University Press, 1997).

Bowie, F., *The Anthropology of Religion* (Oxford, Blackwell, 2000).

Brubaker, R., *Ethnicity Without Groups* (Harvard University Press, 2004).

Cohen, A., *Belonging: Identity & Social Organisation in British Rural Cultures* (Manchester: Manchester University Press, 1982).

Cohen, A., *The Symbolic Construction of Community* (London: Routledge, 1985).

de Coppet, D., ed., *Understanding Ritual* (London: Routledge, 1992).

Farquahar, A., Email to Beltane Fire Society Members (15 February 2007).

Farquahar, A., Talk to Processional Drummers (15 March 2009).

Gray, J., 'Community as Place-making: Ram Auctions in the Scottish Borderland', in V. Amit, ed., *Realizing Community: Concepts, Social Relationships and Sentiment* (London & New York: Routledge, 2002).

Hodkinson, P., *Goth: Identity, Style & Subculture* (Oxford & New York: Berg, 2002).

Howell, S., 'Community Beyond Place: Adoptive Families in Norway', in V. Amit, ed., *Realizing Community: Concepts, Social Relationships and Sentiments* (London & New York: Routledge, 2002).

Hutton, R., *The Pagan Religions of the Ancient British Isles: Their Nature and Legacy* (Oxford: Blackwell, 1991).

Hutton, R., *The Stations of The Sun: A History of the Ritual Year in Britain* (Oxford: Oxford University Press, 1996).

James, E.O., *Seasonal Feast & Festivals* (London: Thames & Hudson, 1961).

Jenkins, R., *Social Identity* (London & New York: Routledge, 2004).

Kendrick, T.D., *The Druids* (Senate, 1927).

Lewis, G., *Day of Shining Red: An Essay in Understanding Ritual* (Cambridge: Cambridge University Press, 1980).

MacCulloch, J.A., *The Religion of the Ancient Celts* (London: Constable, 1911).

McNeill, F.M., *The Silver Bough, Volume 2: A Calendar of Scottish National Festivals: Candlemas to Harvest Home* (Glasgow: William MacLellan, 1959).

McNeill, F.M., *The Silver Bough, Volume 4: The Local Festivals of Scotland* (Glasgow: Stuart Titles Ltd., 1968).

Matthews, J., ed., *The Druid Source Book* (Blandford, 1996).

Ross, A., *Pagan Celtic Britain: Studies in Iconography and Traditions* (London: Routledge and Kegan Paul, 1967).

Schechner, R., *Performance Theory* (New York & London, Routledge, 1994).

Turner, V., *The Ritual Process: Structure and Anti-structure* (New York: Cornell University Press, 1991).

Van Gennep, A., *The Rite of Passage* (London: Routledge & Kegan Paul, 1960).

Webster, G., *The British Celts and their God under Rome* (London: B.T. Batsford Ltd., 1986).

The Traditional Festivals of Northern Europe

Ronald Hutton

Academic specialists in the history of calendar customs have tended, naturally enough, to make case studies of specific examples, often confined to particular communities or districts. What is attempted here is a rarer and broader enterprise: to reconstruct the basic shape of the ritual year across ancient Britain and Ireland, and to suggest any enduring patterns in its celebration.

Modern Pagans, in Europe and America, celebrate a fairly standard ritual calendar of eight festivals formed by interlocking the solstices and equinoxes with the four quarter days that traditionally began the seasons in the British Isles, and which open the Roman months of November, February, May and August. It has a very precise point of origin, in early 1958, when it was created by the most important and influential component group, or 'coven', of the religion of pagan witchcraft, or 'Wicca', which had first appeared, in England, towards the end of the 1940s. Until then, Wiccans had celebrated the quarter days as their main festivals, simply because of the work of Margaret Murray, the leading contemporary proponent of the theory (since discredited) that the people prosecuted for the alleged crime of witchcraft in medieval and early modern Europe had been practitioners of a surviving pagan religion. On the basis of one confession attributed to a Scottish victim of the trials, Murray had declared that the quarter days had been the great feasts of that religion, and Wicca had in part modelled itself upon her imagined reconstruction of it.[1]

What that coven decided in 1958 was to combine these festivals in a ritual calendar with the solstices and equinoxes, giving each of the eight feasts concerned equal status. This was in part because of the traditional importance of midwinter and midsummer as points of festivity and sanctity, but almost certainly because of the prominence of these cardinal points of the sun in the ritual calendar of modern British Druidry. Especially as represented by the activities of the Circle of the Universal Bond, at Stonehenge at the summer solstice and on Tower Hill at the equinoxes, Druids embodied the other most prominent tradition of 'alternative'

1 M.A. Murray, *The Witch-Cult in Western Europe* (Oxford: Oxford University Press, 1921), pp. 109-11.

spirituality, to draw on ancient models, in the Britain of the 1950s. The four solar feasts that they observed had in fact been identified as major ancient Druidic festivals, with no real corroboration from ancient texts, by the Welsh scholar, political radical and forger, Edward Williams, in the 1790s. The meshing of the two separate groups of four festivals — spaced slightly unevenly around the year as one was solar and the other lunar — created the standard Wiccan festival calendar. As Wicca became the template for modern Paganism in general, it was adopted by Pagan groups all over the western world during the following thirty years.[2]

Naturally enough, within a relatively short time, many of the Pagans concerned had become completely unaware that the cycle concerned was a very recent creation, and assumed that it represented the standard one of ancient European pagans. By 1989 it had become to one very influential writer, Caitlín Matthews, who served the rapidly burgeoning constituency of modern Celtic spirituality by providing it with what she took to be ancient and medieval tradition, simply 'the wheel of the Celtic year'.[3] More striking, it was also absorbed by some influential academic writers, with the same assumption. One of these was Alexander Thom, the Professor of Engineering who claimed during the 1960s to have found accurate alignments on heavenly bodies at most prehistoric megalithic sites: he believed that the standard calendar of the British Neolithic and Bronze Age consisted of the eight modern pagan festivals, and duly claimed to have detected alignments built into stone circles and rows from the period which reflected the movements of the sun at each one.[4] Some of his followers maintain a belief in these to this very day.[5]

It may be a worthwhile exercise, therefore, to attempt to lay out, in a single short piece of work, what the ancient festival calendar of ancient northern and western Europe seems to have been, according to the surviving records. To do so means to draw heavily on research that I have published already; but as the work concerned is my own, and as it has not been used for this concise and specific exercise before, such a usage of it may be justified.[6]

One initial observation which I had cause to make when carrying out the original collection of material was that in this field, as in so many others, we are still living in the shadow of the Victorians. It was the great pioneering scholars of the nineteenth century who established what became the

---

2 Ronald Hutton, 'Modern Pagan Festivals: A Study in the Nature of Tradition', *Folklore*, (2008), Vol. 119, pp. 251-60.

3 Caitlin Matthews, *The Elements of the Celtic Tradition* (Shaftesbury: Element, 1989), passim.

4 Alexander Thom, *Megalithic Sites in Britain* (Oxford: Oxford University Press, 1967).

5 E.g. Euan MacKie, 'The Prehistoric Solar Calendar: An Out-of-Fashion Idea Revisited with New Evidence', *Time and Mind*, (2009), Vol. 2(1), pp. 9-46.

6 Collected in Ronald Hutton, *The Stations of the Sun: A History of the Ritual Year in Britain* (Oxford: OUP, 1996), [hereafter Hutton, *The Stations of the Sun*].

standard view of the subject taken by many people in the late twentieth century, and only now is this starting to be revised. In many respects it needs no such revision, because the work by the mighty pioneers was carried out so well. In others, however, a reworking is urgently necessary, because their influence has been such that even straightforward mistakes that they made at moments have been repeated as fact ever since their time. One example of this is the term 'fire festivals', commonly used to describe the quarter days that open the seasons. This derives directly from a book published in 1872 by the folklorist Charles Hardwick, and is based on a misreading of one passage of an early Irish text, *Sanas Chormaic*.[7] This described a fire ritual, to which we shall return, carried out by Irish Druids at the opening of summer; Hardwick misunderstood it to mean that they had enacted it at the beginning of every season, and so decided that each of the festivals that fell at those points of the year had been associated with fire. This simple mistake was then repeated uncritically for a hundred years: in fact, three of the four quarter days had no widespread non-Christian fire rituals connected to them, while the ancient European fire festival *par excellence* was a solar one, that of the summer solstice.

A more extended case-study of the same effect in action concerns the identification of All Saints' Day as the ancient Celtic New Year, and Hallowe'en as its eve.[8] This was first made in 1886, by the leading Oxford philologist Sir John Rhŷs. He did not base it upon any historical research, but upon contemporary folk customs associated with the festival, which he felt were full of references to new beginnings. His political subtext was a desire, as a Celtic nationalist, to create a uniform pan-Celtic culture, as different from those of other Europeans as possible, including its marking of the year. His idea was given much greater currency when it was taken up by the famous Cambridge scholar Sir James Frazer. Initially, Frazer had seen the weakness of it, but he changed his mind when he found it convenient as a prop for a pet theory of his own: that Hallowe'en had been the ancient Celtic feast of the dead. This was based on three arguments: that it actually had been a festival that honoured the dead in the medieval Christian Church; that many medieval Christian festivals had pagan predecessors; and that in other cultures the dead were sometimes honoured at the New Year; which is where Rhŷs's theory became seriously useful. No investigation of the matter was carried out into actual historical records, and large numbers of people just read Frazer's famous book, *The Golden Bough*, and accepted the concept of Hallowe'en as both the ancient pagan feast of the dead and the New Year.

The reality seems to be that that there is no solid evidence for either idea. The early medieval Irish, like the ancient Romans, honoured their dead in April or May, as part of the annual process of cleaning out the home

7 C. Hardwick, *Traditions, Superstitions and Folk-Lore* (Manchester, 1872), pp. 30-40.
8 Hutton, *The Stations of the Sun*, pp. 363-65.

after the winter had gone and putting the memory of its sufferings (in the course of which most natural deaths would have occurred) behind them. The idea of doing so at the beginning of November appeared in the network of monasteries based on the French mother house of Cluny, in the eleventh and twelfth centuries, as part of the developing doctrine of Purgatory, by which most human beings would be prepared for heaven by a period of punishment in which they would atone for their sins in life. The custom that it embodied was for the living to pray to saints to intercede for their dead friends and relatives, to shorten their sufferings, and that is why the first two days of November became dedicated to All Saints and All Souls. It is not clear why the new holy days were placed at the beginning of winter — perhaps simply because, as that was normally the most lethal of all seasons, the attention of people might then be best concentrated on their fate after death. None of this was known to Frazer, who was not a historian, and historians in fact took no interest in the matter. Instead, the construction made by Rhŷs and Frazer was repeated, as fact, in book after book produced by prominent Celticists and archaeologists, such as Anne Ross, Nora Chawick, Lloyd Laing and Graham Webster, until the end of the twentieth century.[9]

There is good evidence that winter was regarded as the first of the four seasons in medieval Ireland — it is represented as such, for example, in the twelfth-century text, *Tochmarc Emire*. This is not, however, the same as reckoning the opening of the calendar year from its commencement.

These are a couple of examples of the checking and amendment that needs to be carried out now to the achievements of the Victorian pioneers of research into the ritual year. When all the actual historical sources for the subject are pooled, a coherent picture of an ancient ritual calendar does start to emerge, and that will be laid out now. The first festival of the year, in reckoning of dates, was also the greatest by far in terms of length and intensity of celebration, attracting to it, for example, a quarter of all the seasonal customs ever observed in the island of Britain. This was the one based on the winter solstice, which marked the New Year for the pagan Romans, Germans, Scandinavians and Celtic-speaking peoples. It was known to the Anglo-Saxons as Modranicht, the 'Mother Night', to the Scandinavians as Yule, while by historical times speakers of Celtic languages employed a range of terms for it based on the Christian 'nativity', such as the Irish Nollaig and the Cornish Nadelek. In modern times it is possible to pinpoint the moment at which the sun technically makes its turn towards the opposing tropic. Until the nineteenth century, however, instead of rising and setting at visibly slightly different points on the horizon each day, as it did for most of the year, at midwinter and midsummer it entered a period at which it appeared, to the eye, to do so at the same points, its northerly or southerly maximum. This time, of four to five days, was a

9 Hutton, *The Stations of the Sun*, pp. 408-11.

distinct, dramatic and numinous one at the extremes of the year, and the word 'solstice' itself derives from the Latin for 'the sun stands still'. The ancient Romans put their celebrations on either side of it, with one great festival, Saturnalia, before it began, and the other, the Kalendae, on the far side, to mark the New Year as the days were visibly growing in length again.[10]

The common rituals of the feast can be divided according to functional categories. The first category of them was concerned with keeping cheerful in the most cheerless time of year.[11] It was one of want, in nature, and so humans would feast and make merry. It was dark and cold, and so they would light and heat their homes with special candles and log fires. The trees were mostly bare of greenery, and so people would decorate their homes and holy places with branches of the few that were still carrying foliage, as symbols of enduring life — the most common were holly and ivy. The second category of activities consisted of expressions of hope and fear for the coming year, in a great variety of rites of purification and blessing. In Britain alone, for example, the following kinds were used. The Gaelic-speaking Scots would 'sain', or exorcise their homes and animals with burning juniper, while the southern English 'wassailed' their animals, fields and orchards by singing blessings to them. The northern English and Lowland Scots placed much more emphasis on the first person to set foot in the home after the coming of the New Year, and the manner of her or his coming.[12]

The third category consists of misrule. If midwinter was one of the most depressing and restrictive points of the year in many respects, this feature actually made human beings safest from each other — the condition of the roads and seas, and the lack of warmth and daylight, made military operations exceptionally difficult, and communities most secure from disruption by outsiders. As a result, they were able to relax and loosen the usual social bounds in a manner not possible at other seasons. The Romans had the custom of the Saturnalia King, who devised pranks and games for a group of friends, while masters and mistresses waited upon their servants at table. The medieval and early modern periods across Europe had the Lord of Misrule, or Bean King, chosen for the period of the midwinter festival to preside over wealthy households and dedicate them to fun. Cathedrals established a Feast of Fools for junior clergy, increasingly replaced in the later Middle Ages by the election of one of the choirboys to take the place of the actual bishop and carry out most of his functions during the same period. English schoolchildren between the sixteenth and nineteenth

10 Hutton, *The Stations of the Sun*, pp. 1-8.
11 Hutton, *The Stations of the Sun*, pp. 34-41.
12 Hutton, *The Stations of the Sun*, pp. 42-53.

centuries were commonly permitted to fortify their classroom against their teacher, and if they could hold out were awarded an extra day of holiday.[13]

The fourth category of customs was devoted to charity, and the recognition that the social cohesion of a community would be badly damaged if some of its members starved while the rest feasted and made merry. From the time that records for such activities begin in the late Middle Ages, wealthier members of European villages and towns made donations to enable the old, infirm and very young to have something to eat or drink at this time. The able-bodied poor, from earliest times, earned the same benefits by visiting the houses of wealthier neighbours, or local inns and alehouses, to perform songs and plays for the entertainment of the inhabitants. This activity neatly supplied a dual function: of providing diversions for the comfortably-off, at no effort to themselves, and enabling the needy to earn hospitality, and, in later times, money, without the shame of begging.[14]

The next important group of festivals occurred around the beginning of February, and signalled the opening of spring. This was a much less straightforward and universal phenomenon to celebrate than midwinter: solstices occur all over the world, but springtime is relative to the local climate. There is no sign of this set of feasts in Scandinavia or north-eastern Europe, for example, where winter conditions remain dominant into March and April. Further south things were different: in the Mediterranean the February weather is usually clement compared with that of midwinter. Even on the latitude of Britain, February is traditionally the month which, though still freezing cold, drives the darkness from the afternoon and brings the first buds and flowers to the land, and the first love songs of birds. That is why a major clutch of celebrations and rites are recorded around its opening. In early Ireland, the feast that heralded spring, and was put on 1 February, was called Imbolc or Oimelc, a name which has so far defied the explanations of philologists, while records of no actual rituals have survived for it. The Anglo-Saxon equivalent was Solmonath, which the early English historian Bede explained as meaning 'cake-month', because people made cakes and offered them to deities at this time.[15]

Pagan observations of the season are heavily overlaid in the existing information by the medieval Christian equivalents. The Irish converted Imbolc into the annual feast of their mother saint, Bridget, whose cult spread from Leinster to all the lands in which Irish influence was strong in the Middle Ages, including the Hebrides and Isle of Man (but not to other Celtic societies where it was weaker, such as Wales and the Scottish Highlands). The common belief of this cultural zone was that Bridget would visit households at this time and bless them if ceremonies were observed to

13 Hutton, *The Stations of the Sun*, pp. 95-111.

14 Hutton, *The Stations of the Sun*, pp. 9-24, 54-94.

15 Hutton, *The Stations of the Sun*, pp. 134, 140.

welcome her.[16] Everywhere else in Europe, a much more general festival of a female saint was observed instead – that of the Purification of the Virgin Mary.[17] This was inspired by a specific episode in the New Testament in which the infant Christ was taken by his parents to the temple at Jerusalem, so that Mary could purify herself of childbirth, forty days after the event, according to Jewish religious custom. There, according to the tale, he was recognised by an old man called Simeon as the Messiah of Israel and 'a light to lighten the Gentiles'.

The theological and political significance of this passage was profound: it appeared to provide proof that Jesus had been acknowledged, from the beginning, as having come to redeem all humanity instead of merely his own people, the Jews. As such, a festival to highlight and celebrate it was instituted in Rome in the seventh century – on 2 February, as that fell forty days after the Nativity – and spread rapidly northward from there. As it did so, however, it changed character. In the Mediterranean lands, it was the aspect of Mary's pilgrimage as an act of purification which was emphasised, meshing with the ancient Roman associations of February. The very word *februa* signifies cleansing, and the season was one of preparations and renewals for the activities of the coming year, now that winter was over.

North of the Alps, the cold weather normally still lingered, and there the association of the Biblical passage with light became most prominent. The central custom of the new feast became the blessing and kindling of candles, with a liturgy that emphasised the power of light to drive back darkness, especially therapeutic at this time of year. The consecrated candles themselves became magical objects, commonly believed to ward off fire from the owners' homes, at a time when almost all dwellings were made of wood. They gave the festival its characteristic English name of Candlemas (the Welsh equivalent being 'Gwyl Fair', the Feast of the Virgin). It is possible that the opening-of-spring rites had been associated with female deities or with fire before Christian times, but there is absolutely no evidence for this.

After this, there is no sign of any clustering of feasts and rites for two months, and certainly not around the spring equinox. The time was the one of greatest want, the stocks of food amassed for winter now being at their lowest, and the work of ploughing, sowing and caring for young livestock at its hardest. According to the Julian calendar, the Romans marked the opening of the official year at the equinox, when the new office-holders took up their posts: this was most important in the case of military commanders, because the campaigning season was now opening. The popular New Year, however, remained at the end of the winter solstice, and there was no special clustering of Roman religious festivals around 21

16 Hutton, *The Stations of the Sun*, pp. 134-38.
17 Hutton, *The Stations of the Sun*, pp. 139-45.

March.[18] It was Christianity that gave a new significance to the equinox, by tying the date of its greatest festival, of Easter, to it, following the Jewish dating of the feast of the Passover. Even so, Easter could fall almost a month after the equinox had passed. Nor, while there is ample evidence that prehistoric monuments in north-western Europe were aligned on the solstices, is there equivalent agreement among archaeoastronomers that there are clear alignments on the equinoctial sunrises and sunsets.[19]

By contrast, there is no mistaking the presence and importance of the next major cluster of seasonal celebrations, those which hailed the coming of summer and were placed in late April or early May with the general adoption of the Roman calendar. To the Norse, these were called the Summer Nights, to the Welsh, Calan Mai, and to the English, May Day. The Irish sometimes called their equivalent Cetsoman, and sometimes Beltine or Bealtainne, though properly speaking the latter pair of names indicated not so much a festival as a ritual, signifying 'holy fire' or 'lucky fire'. This represented the Irish expression for a ceremony found right across northern Europe, wherever people lived, as they did in early Ireland, with an economy dependent on the moving of livestock from winter to summer pastures. For some, this merely meant taking them from stables, folds, byres and infields onto open pastures and commons on the edge of the community, while for others it could involve driving them for miles into uplands and living with them there. The rite consisted of kindling and blessing two fires and driving the animals between them, or a single one and moving the stock around it; sometimes the people walked between the fires or jumped over them instead. The consecrated fire was believed to give the herds and flocks, and their human minders, protection against the human and animal predators, and diseases, of the summer pastures. The use of it was certainly ancient, not merely because it was so widespread but because an Irish text written around the year 900, *Sanas Chormaic*, states specifically that it was inherited there from pagan times, when it was carried on by Druids. It was not specifically Celtic, however, being found all over northern Europe where people moved groups of farm animals between seasonal pastures.[20]

In regions where arable farming was dominant, and humans depended for survival on growing and harvesting crops, the holy fires were not lit, and another custom obtained instead. This was the gathering of garlands of greenery and flowers, and the carrying of them in procession and decking of homes and holy places with them. Very often, a wooden pole would be erected and decorated with them, and the people would dance around it.

18 Hutton, *The Stations of the Sun*, p. 2; H.H. Scullard, *Festivals and Ceremonies of the Roman Republic* (London: Thames and Hudson, 1981), pp. 84-95, 182-88.
19 Clive Ruggles, *Astronomy in Prehistoric Britain and Ireland* (London: Yale University Press, 1999), passim.
20 Hutton, *The Stations of the Sun*, pp. 218-25.

The English expression for this is the may-pole, and ever since Freud wrote (and indeed, for the historically-minded, ever since English Puritans condemned the custom in the seventeenth century), many modern people have felt that they know what it represents. In fact, there is not the slightest of evidence that the medieval and early modern people who erected may-poles saw anything phallic in them, and indeed it would have been hard to do so, as the stark outline of the pole would have been softened with coverings of greenery and blossoms. The ribbons that are now attached are a Sicilian custom, which did not reach Britain until they were made popular in the London theatre in the nineteenth century. The pole instead represented a super-tree, emblematic of, and displaying, everything sprouting and flowering in the district at the time — a celebration and honouring of the power of vegetation.[21]

In modern tradition, the start-of-summer festivals have acquired especial connotations of love and sexuality; but there is no older support for this idea. At the turn of April and May, the weather is still just too cold across most of northern Europe for outdoor activities to be comfortable, unless both well clad and vigorous. The season of open-air village feasts and sports began in Britain towards the end of May, after the Christian feast of Whitsun, while a good index of sexual activity can be provided by looking at the dates of baptisms in Tudor and Stuart English parish registers. Both in and out of marriage, the conception of children rises steeply towards the end of May and peaks in June and July, when the ditches have become dry, the leaves thick and the grass and crops tall; all vital considerations in an age of minimal privacy in the average home. This makes the sexiest major festival of the year in old northern Europe not that which opened summer but that which celebrated its apex, at the solstice.

All across Europe, midsummer was celebrated energetically, and in two ways. One was with more customs in which people gathered, paraded, fixed up and danced round foliage and flowers, now at the fullness of their power and beauty. The other was to bless and kindle fire again — and this time the custom was found in every economic region, from Norway to Algeria and Ireland to Russia, making midsummer the greatest fire festival of the ancient world.[22] This may have been related in some measure to sun-worship, at a time when the sun had its greatest dominion over night. The actual content of the rites, however, points to a different purpose, of protection. Once again, the primeval utility of fire — to drive away darkness, cold and predatory beasts — was honoured symbolically, as people danced around bonfires and jumped over them and carried torches around their fields of crops and herds and flocks of beasts. This activity was propelled by an aspect of late summer and early autumn that has now been almost completely forgotten: however delightful a season it may be in natural

21 Hutton, *The Stations of the Sun*, pp. 226-43.
22 Hutton, *The Stations of the Sun*, pp. 311-21.

terms, in human terms it was one of the most dangerous. With the land so dry, warm and open, the rivers so low and the seas so calm, it was the favourite time of year for armies, fleets, pirates and rustlers to operate. Out in the summer pastures, livestock was at its fattest and most tempting to predators of all kinds, and also most likely to catch epidemic diseases. The crops were now grown enough to be destroyed, completely, by sudden storms or long periods of rain, or blighted by diseases of their own. It is also the season, as every modern pet-owner knows, when insects most multiply in numbers, and the fleas of medieval and early modern Europe carried something much deadlier than irritation in their bite: they brought bubonic plague, which always struck from mid June onwards, peaking in August and September. Those who lit the protective fires had a long list of misfortunes against which the magic of the flames was hoped to be effective.

After this, another six weeks passed before Europeans locked into the cycle of marking seasonal openings once more, with the beginning of autumn. This was placed around the beginning of August, for two good economic reasons. For those engaged in pastoral farming, it fell halfway through the season of the summer pastures, a good time for games and celebrations to be held by those who needed some festivity and for whom the open season had gone well thus far; those in need of further reassurance would use it for an additional honouring of divine powers.

In arable regions, August was the main cereal harvest month until modern methods enabled the process to be extended forward and backward in the twentieth century. The work of gathering in the grain crops was the most back-breaking and nerve-racking of the entire year, and was customarily opened by the ritual cutting of the first ears and the offering up of them in temples or churches in the hope of good weather and so relatively easy and profitable labour. In medieval England the festival accordingly became known as the loaf-mass, or Lammas, as these first fruits were baked into a sacred bread; in Ireland it was called Lughnasadh, which could derive from the name of the pagan god Lugh, the many-skilled, or from the word for law and justice, indicating that tribunals were held then to settle disputes that had arisen in the course of the summer.[23]

There was no general point at which the main economic activity of the year came to an end, and so no general set of festivals between early August and late October, because local economies were so varied. As the cereal harvest ended on each farm, the proprietors would hold the best possible meal and party for all who had helped them gather it in; these harvest suppers, held throughout August in north-western Europe, are recorded back to Roman times.[24] Other kinds of harvest, such as of vegetables and fruit, ended in smaller but similar celebrations. Communities which depended on fishing and sailing tended to mark the end of their main

23 Hutton, *The Stations of the Sun*, pp. 327-31.
24 Hutton, *The Stations of the Sun*, pp. 332-47.

season of activity near the end of September, at the medieval Christian feast of Michaelmas. Last of all, those which depended on herding would bring back their animals from the summer pastures in the course of October, as the grass ceased to grow and the frosts began to fall. By the end of that month, absolutely everybody in a traditional community would be expected to have come home, and it was around that time that the final general seasonal cluster of ritual and celebration took place. It was known in Scandinavia as the Winter Nights, to the Irish as Samhain (probably meaning summer's end), to the Welsh as Calan Gaeaf (the beginning of winter), and to the Anglo-Saxons as Blodmonath, the blood month. The Christian equivalents are the feasts of All Saints and All Souls, which have given us the modern name of Hallowe'en.

The rites and traditions associated with it suggest not so much a time for honouring the dead as for confronting the prospect of one's own death, at the beginning of what was in nature the most unpleasant of all seasons. At best it would be cold, dark and restrictive; at worst, lethal. The customs associated with it therefore fall into three categories. One consists of an acknowledgement, and commonly a mockery, of the powers of dark and cold. The second focused on divination to see how bad the winter was likely to be, and in particular who was likely to die in the course of it. The third consisted of feasting and party games, as a means of raising morale before the trials to come set in. Feasting would have been particularly natural, because of the annual need to slaughter those animals which could not be stabled and fed through the coming cold months; this is the economic reality behind the Anglo-Saxon name for the period. As this would also be the definitive season at which all the warriors, traders, pilgrims and members of every kind of farming community would have returned to the home base, it was also a natural time at which to swap tales, take stock, compare notes and draw lessons concerning the active time of the year past.[25]

This gives us a basic cycle of six general festivals, or clusters of festivity and ritual, which correspond well to six of those in the modern Pagan list; the equinoxes are those missing, and seem to represent the point at which Edward Williams' list of feasts lost touch with ancient and medieval reality. It may be coincidental, or indeed a false observation, but I have noted that when wheels are represented in the art of Roman Britain and Gaul to associate solar or celestial powers with deities, they seem usually to have six spokes.[26] This is the best visual evidence available for any kind of a sense of cosmic cycle in ancient north-western Europe.

25 Hutton, *The Stations of the Sun*, pp. 360-70.

26 Cf Anne Ross, *Pagan Celtic Britain* (London: Routledge, 1967); Miranda Green, *The Gods of the Celts* (Gloucester: Sutton, 1986); and Graham Webster, *The British Celts and their Gods under Rome* (London: Batsford, 1986).

I am more concerned here, however, with the fate of the traditional festive calendar in modern times. In the course of the last two hundred years, the traditional reckoning of the seasons in Britain has largely, and in official terms wholly, been replaced by the American system of opening them at the solstices and equinoxes: so that, for example, winter formally begins at the winter solstice. This actually does correspond well to the climate and ecology of at least the northern United States, although less well to traditional British farming cycles. Its influence may be attributed in large part to the profound impact of American ideas, especially on the British but also on the western world as whole; but also to a much deeper and more important alteration in human relationships with the natural world. Even in Britain, the American reckoning corresponds well to climate: the coldest part of the year is commonly between December and March, the hottest from June to September, between March and June the land becomes green, and between September and December the leaves change colour and fall. Although it makes nonsense of older surviving terms such as 'midwinter' and 'midsummer', the new reckoning suits well enough a society in which most people are completely cut off from the realities of the farming cycle.

What has happened in that society is that seasonal festivals have been reordered into a new system, broadly similar from country to country, but with distinctive national ingredients. In Britain it commences at Hallowe'en when children are encouraged to mock and impersonate spirits of the dark while soliciting presents from neighbours. The essence of the custom, which is an American one based ultimately on Irish celebrations of the feast of All Saints, seems to be to enable them to come to terms with traditional images of fear.[27] It is followed by the new fire festival of Guy Fawkes's Night, developed in the seventeenth century as a Protestant celebration to mock Catholicism, and retained as it provides merriment at the opening of winter.[28]

These feasts open a two-month period of activity, with town carnivals in November and an increasing atmosphere of excitement and preparation in December, which culminates in the festive season par excellence at the winter solstice. This formally opens with Christmas, retaining all the key features of the traditional midwinter festival but redefined as pre-eminently a celebration of the family, centred on children; a function embodied in its spiritual patron, the American figure of Santa Claus who was born in New York in 1822. Like the ancient Romans, the modern British have a twin-peaked midwinter festival, as the family-centred Christmas is commonly followed (and in some measure balanced) by the chance that

27 Hutton, *The Stations of the Sun*, pp. 379-85.
28 Hutton, *The Stations of the Sun*, pp. 393-407.

most people have to celebrate New Year in adult fashion with companions of their choice.[29]

After that comes a series of spring festivals dedicated to different aspects of the human life-cycle: Valentine's Day for lovers, Mothering Sunday for mothers and Easter providing a second focus in the year on children and their tastes. Both of the first two customs are embedded in British tradition, but both had effectively died out in the early twentieth century and were revived in a new, general, and commercialised form because of American influence.[30] The 'ritual half' of the modern year therefore spans the period between the end of October and that of April. The other half is now that in which most people take their annual holidays in the new sense, of a period of leisure and recreation without any religious connotations. The nature of these has exactly reversed in the period between 1850 and 1950. Before that time, pleasures taken as a release from work were mainly enjoyed within a person's own community, in the form of wakes, ales, sports days, harvest festivals and dances. Since then, a large part of the point of a true holiday has been to get completely away from that community, whether defined as the household, the workplace or the neighbourhood. It has become one of the points of the affirmation of an individual, or family-centred, identity against those imposed by the external factors of residence and labour.

These major alterations, which have effectively done away with much of the traditional festive calendar, are due in part to specific alterations in technology and society, and have been associated with a degree of American influence on British customs which is often underestimated. The culture of the United States, however, has only been so influential because it has recently led the world in promoting and adapting to shifts of lifestyle and perception which have become very widespread. One, above all, has had the crucial impact on the nature of the ritual calendar: urbanisation and industrialisation. The ancient calendar was, after all, founded on a way of life which continued in essential respects unaltered from ancient times until the nineteenth century — one in which the bulk of the population was directly or indirectly concerned with farming processes, leavened by trade and war. The revolution in attitudes to the seasonal cycle, since about 1850, has reflected one single tremendous cosmological alteration: the displacement of the natural world from the centre of the western mental universe, and the substitution of human identities and needs. That is why the old ritual calendar marked and celebrated the processes of nature, and the new one is devoted instead to humanity.

---

29 Hutton, *The Stations of the Sun*, pp. 112-23.
30 Hutton, *The Stations of the Sun*, pp. 146-68, 198-203.

Bibliography

Green, Miranda, *The Gods of the Celts* (Gloucester: Sutton, 1986).

Hardwick, Charles, *Traditions, Superstitions and Folk-Lore* (Manchester, 1872).

Hutton, Ronald, *The Stations of the Sun: A History of the Ritual Year in Britain* (Oxford: Oxford University Press, 1996).

Hutton, Ronald, 'Modern Pagan Festivals: A Study in the Nature of Tradition', *Folklore,* (2008), Vol. 119, pp. 251-73.

MacKie, Euan, 'The Prehistoric Solar Calendar: An Out-of-Fashion Idea Revisited with New Evidence', *Time and Mind,* (2009), Vol. 2(1), pp. 9-46.

Matthew, Caitlín, *The Elements of the Celtic Tradition* (Shaftesbury: Element, 1989).

Murray, Margaret Alice, *The Witch-Cult in Western Europe* (Oxford: Oxford University Press, 1921).

Ross, Anne, *Pagan Celtic Britain* (London: Routledge, 1967).

Ruggles, Clive, *Astronomy in Prehistoric Britain and Ireland* (London: Yale University Press, 1999).

Scullard, H.H., *Festivals and Ceremonies of the Roman Republic* (London: Thames and Hudson, 1981).

Thom, Alexander, *Megalithic Sites in Britain* (Oxford: Oxford University Press, 1967).

Webster, Graham, *The British Celts and their Gods under Rome* (London: Batsford, 1986).

Cyberspace and the Sacred Sky

Frances Clynes

The sacredness of the celestial realm found expression the works of Plato, in Gnosticism and in Christianity. Today, Internet terminology is packed with celestial terms such as clusters, and parallels with celestial terms from the Deep Internet to Cloud Computing. Just as the sky has been viewed as sacred, cyberspace theorists write of cyberspace as a sacred space, and it has been compared to Eliade's sacred space and Cassirer's mythical space. This talk looks at current views of cyberspace and asks if they are a re-packaging of the age-old concept of a sacred sky?

In the twenty-first century, computing terminology, particularly network and Internet terminology, is packed with terms with celestial connotations. Groups of network nodes are known as clusters.[1] The symbol for the Internet is a cloud, and Cloud Computing, a separate term from the cloud symbol, is the current buzzword.[2] Descriptions of the Dark Internet or Deep Internet appear similar to those of dark matter and deep space.[3] For example, according to NASA, ninety-six per cent of the universe is composed of invisible dark energy and dark matter.[4]

Similarly, the Dark Internet, so called because it is invisible on the surface, is estimated to be at least 500 times greater than the surface accessible Internet.[5] The tendency to give Internet technology names already associated with the sky suggests that in some way the Internet appears, at least to some people, to share some common characteristics with the celestial realm. This in turn suggests that humanity's relationship with the sky and space is being repeated in its relationship with the Internet or cyberspace.

---

1 George Aggelou, *Mobile Ad-hoc Networks* (New York: McGraw-Hill, 2005), p. 120.

2 Anon., *PC Magazine Encyclopedia*, http://www.pcmag.com/encyclopedia_term/0,2542,t=cloud&i=39847,00.asp and http://www.pcmag.com/encyclopedia_term/0,2542,t=cloud+computing&i=57964,00.asp [accessed 7 July 2009]

3 Michael K. Bergman, 'The Deep Web: Surfacing Hidden Value', *The Journal of Electronic Publishing* (August 2001), Vol. 7(1), Issue title: *Taking License: Recognizing a Need to Change*, available at http://quod.lib.umich.edu/cgi/t/text/text-idx?c=jep;view=text;rgn=main;idno=3336451.0007.104 [accessed 7 July 2009], [hereafter Bergman, 'The Deep Web'].

4 Charlie Plain, 'A Dim Light Shines on Dark Matter', NASA website, http://www.nasa.gov/missions/deepspace/chandra_dark_matter_halo.html [accessed 7 July 2009].

5 Bergman 'The Deep Web'.

The Internet came into being in the Unites States of America. According to John Naughton in his history of the Internet, the spark that lit the fuse that brought about the Internet's birth was the Soviet launch into space of Sputnik, the world's first artificial satellite, in October 1957. On that day, Naughton claims, there was a strong reaction in America, or, to use his words, 'the US went apeshit'.[6] He states that it was in an effort to end rivalries between different arms of the US military, which it was believed had resulted in the US falling behind in aerospace technology, that the Advanced Research Projects Agency (ARPA) was set up in 1958. Although their brief with regard to space would be handed over to NASA later that year, it was ARPA who pioneered the world's first computer network, initially called ARPANET. The original purpose of ARPANET was to allow ARPA-sponsored researchers in various locations to collaborate on research projects.[7] The first permanent ARPANET link was established between computers at University College Los Angeles (UCLA) and at the Stanford Research Institute in 1969. By December 1969 a four-node network, including two more computers at University College Santa Barbara and the University of Utah, was connected. By 1981 there were 213 nodes in different parts of the world. In 1983 the US military portion of the ARPANET was broken off as a separate network. Further technical developments such as the development of Internet Protocols and a global addressing system allowed the creation of an open architectural environment. Widespread use of computers and a drop in prices allowed it to flourish.[8]

The Internet today is a small global network of computer networks. A crucial part of its design is that each network should be able to work on its own, developing its own applications without restraint and requiring no modification to participate in the Internet. The failure of one network should not affect the others. In 1991, the World Wide Web was released to the public. This is a system that allows links which are hidden behind text to retrieve documents or pages using HyperText Markup Language (HTML). This language made the system user-friendly and accessible to people unfamiliar with complex technology.[9] The web's spread was rapid and exponential. By 1990 there were 2500 host networks throughout the world (although mainly in developed countries), connected to the Internet. By

6 John Naughton, *A Brief History of the Future: The Origins of the Internet,* second edition (London: Phoenix, 2000 [Weidenfeld & Nicolson, 1999]), [hereafter Naughton, *A Brief History of the Future*], p. 79.

7 Naughton, *A Brief History of the Future*, pp. 131-139.

8 Barry M. Leiner, Vinton G. Cerf, David D. Clark, Robert E. Kahn, Leonard Kleinrock, Daniel C. Lynch, Jon Postel, Larry G. Roberts and Stephen Wolff , 'A Brief History of the Internet', The Internet Society, http://www.isoc.org/internet/history/brief.shtml [accessed 4 July 2009].

9 Anon., 'From Arpanet to World Wide Web', *Internet History*, University of Leiden, http://www.leidenuniv.nl/letteren/internethistory/index.php3-c=5.htm [accessed 4 July 2009].

2000 there were an estimated 361 million users of the Internet. In March 2009 the figure was estimated to be over 1,596 million.[10]

In 1982, while the Internet was still known as ARPANET, a small network connecting researchers, the term cyberspace was coined by American author William Gibson. Gibson first used the word in 1982 in *Burning Chrome*, and developed it in his 1984 novel, *Neuromancer*. In this much-quoted passage from *Neuromancer*, Gibson defines cyberspace:

> Cyberspace. A consensual hallucination experienced daily by billions of legitimate operators ... A graphic representation of data abstracted from the banks of every computer in the human system. Unthinkable complexity. Lines of light ranged in the nonspace of the mind, clusters and constellations of data. Like city lights, receding ...[11]

In *Neuromancer* and the novels that followed which gave birth to the genre that was to be known as cyberpunk, the protagonist, who is almost always male, lives in a world dominated by technology and spends a lot of time in cyberspace. He frequently breaks laws, flouts authority and takes risks. His technical skills lead him into dangerous situations, but usually bring about his escape. Cyberspace is seen as both a place of adventure and excitement and a place where people can become rich.[12]

The term cyberspace became widely used in connection with the Internet during the 1990s, particularly in academic circles. John Perry Barlow is credited with being the first to use the term in this way in his 1990 essay 'Crime and Puzzlement'.[13] In 2006, Karaflogka argued that, although she sees them as entirely distinct from each other, the notion of cyberspace contributed to the growth of the Internet. Even though, as she stated, the Internet of the twenty-first century is 'a far cry' from the cyberspace described by Gibson, the terms are often used interchangeably.[14]

Like Bell before her, Karaflogka differentiates between cyberspace and the physical hardware and software that allows the Internet to exist, or what Bell calls,

10 Anon., *World Internet Statistics*, http://www.internetworldstats.com/stats.htm [accessed 5 July 2009].

11 William Gibson, *Neuromancer* (London: Harper Voyager, 1995), p. 67.

12 For example Case in William Gibson's *Neuromancer*, or Bobby Newmark in William Gibson's *Count Zero*.

13 John Perry Barlow, 'Crime and Puzzlement', on the *John Perry Barlow Library* website, http://w2.eff.org/Misc/Publications/John_Perry_Barlow/HTML/crime_and_puzzlement_1.html [accessed 5 July 2009], hereafter Barlow, 'Crime and Puzzlement'].

14 Anastasia Karaflogka, *E-religion: A Critical Appraisal of Religious Discourse on the World Wide Web* (Equinox: London, 2006), [hereafter Karaflogka, *E-religion*], p. 3.

> the network of computers, modems, communication links, nodes and pathways that connect users into something (or some things) like the Internet, the World Wide Web, the information superhighway and so on.[15]

However, Bell believes there is another dimension to the Internet.[16] He goes on to state,

> We have to move beyond the simple answer that physically we are seated in front of a monitor ... We are simultaneously making ourselves over as data, as bits and bytes, as code, relocating ourselves in the space behind the screen, between screens, everywhere and nowhere.[17]

In Bell's other dimension, the person in front of the monitor is not just visiting a website, they are clearly entering another world. This world is what Karaflogka calls cyberspace, or, as she puts it, 'where the act of connectivity and interactivity takes place'.[18] She describes cyberspace as a place which, although part of the network, is nevertheless completely different from it.[19] To clearly differentiate between the two she uses the term 'Internet' to refer to the physical network and 'cyberspace' to refer to the place Bell calls 'everywhere and nowhere'. This distinction will also be used here.

Cyberspace has been compared to the Christian Heaven by Wertheim, the Gnostic Heaven by Davis, the Heavenly City, the New Jerusalem and the Garden of Eden by Benedikt, and Utopia by Robins.[20] It has been claimed by Stenger that it fits the description of Mircea Eliade's sacred space, in that it meets the conditions necessary for a hierophany or eruption of the sacred into the profane world that results in the territory, in this case cyberspace, becoming detached from its surroundings and becoming qualitatively different.[21] A typical example is the following quote from Nicole Stenger:

---

15 David Bell, 'Cybercultures Reader: A User's Guide', in David Bell and Barbara M. Kennedy, eds., *The Cybercultures Reader* (London: Routledge, 2000), [hereafter Bell, 'Reader'], p. 2.
16 Bell, 'Cybercultures Reader', p. 2.
17 Bell, 'Cybercultures Reader', p. 3.
18 Karaflogka, *E-religion*, p. 24.
19 Karaflogka, *E-religion*, p. 25.
20 Margaret Wertheim, *The Pearly Gates of Cyberspace: A History of Space from Dante to the Internet* (New York and London: Norton, 1999), p. 18; Erik Davis, *Techgnosis: Myth, Magic and Mysticism in the Age of Information* (London: Serpents Tail, 2004). [1998]), [hereafter Davis, *Techgnosis*], p. 97; Michael Benedikt, 'Introduction', in Michael Benedikt, ed., *Cyberspace: First Steps* (Cambridge Massachusetts and London: MIT Press, 1991), p. 15; Kevin Robins, 'Cyberspace and the World We Live In', *Body & Society* (1995), Vol. 1, p. 135.
21 Mircea Eliade, *The Sacred and the Profane: The Nature of Religion*, trans. W.R. Trask (San Diego, New York and London: Harcourt Inc, 1959); Nicole Stenger, 'Mind is a

> On the other side of our data gloves we become creatures of coloured light in motion, pulsing with golden particles ... We will all become angels, and for eternity ... Cyberspace will feel like Paradise, a space for collective restoration of the habit of perfection.[22]

Stenger is French, but has worked in the US since 1992.[23] This fact is significant because, as noted by Karaflogka, most cyberspace theorists are American. As Karaflogka states, not only is the Internet an American invention, but cyberspace as a place is an American concept.[24] One of many possible explanations for the American domination of cyberspace is the importance of the concept of the Western frontier to the American sense of national identity. It is argued that there is a spiritual need for an open frontier in the American psyche and it was in recognition of this that the television programme *Star Trek* named space 'the final frontier'. As the old frontier to the West no longer existed, space offered a new frontier to be explored, and in the 1990s cyberspace offered yet another.[25]

In 1893 Frederick Jackson Turner wrote about a report in the 1890 census in the United States, that since isolated settlements existed throughout the American West, the frontier line was no more. He believed that it marked the end of a great historic movement that had helped form the American view of the world:

> Up to our own day [1893] American history has been in a large degree the history of the colonization of the Great West. The existence of an area of free land, its continuous recession, and the advance of American settlement westward, explain American development.[26]

Turner predicted that the expansive character of American life would not cease and 'the American energy will continually demand a wider field for its exercise'.[27] In an article about the concept of the frontier and its profound impact on the American psyche, Dave Healy attempts to describe how this has occurred. He believes that, as predicted by Turner, the search for new

---

Leaking Rainbow', in M. Benedikt, ed., *Cyberspace: First Steps* (Massachusetts: MIT, 1991), [hereafter Stenger, 'Mind is a Leaking Rainbow'], p. 55.

22 Stenger, Mind is a Leaking Rainbow', p. 52.

23 Diane Thome, 'Reflections on Collaborative Process and Compositional Revolution', *Leonardo Musical Journal*, (1995), Vol. 5, p. 29.

24 Karaflogka, *E-religion*, p. 19.

25 Anon., http://www.startrek.com/startrek/view/series/MOV/005/index.html [accessed 13 July 2009].

26 Frederick Jackson Turner, 'The Significance of the Frontier in American History', *Proceedings of the State Historical Society of Wisconsin, December 14, 1893*, online at http://xroads.virginia.edu/~HYPER/TURNER/ [accessed 5 July 2009], [hereafter Turner, 'The Significance of the Frontier'].

27 Turner, 'The Significance of the Frontier'.

wilderness continued to the 'final frontier' of space and on to the new frontier of cyberspace 'where some of the same tensions that characterized American settlement continue to be played out'.[28] He notes that, like Huck Finn, the console cowboy of cyberspace is a young male who finds salvation in escaping the civilising influence of authority figures. Using his keyboard he charts the unexplored and sometimes dangerous territory of cyberspace.[29]

The land beyond the frontier is seen as a lawless place. Healy describes it as 'that hypothetical border-land between civilisation and the wilderness'.[30] In this he echoes Turner who called it 'the meeting point between savagery and civilization'.[31] Positing the existence of a spiritual need (in America) for an open frontier, and the idea of space travel as an expression of this need, Mary Midgley believes that a certain lawlessness is part of the attraction: 'Symbolically space stands for freedom. Negatively this means not being interfered with by others; positively, it means increased opportunities for action'.[32] She goes on to say that space is heir to the literal, geographical frontier which Americans were for a time accustomed to see as symbolising unlimited freedom, always available to them if they chose to use it.[33] Healy also believes that the idea that space represents freedom has a special resonance for 'a nation that began with an errand into the wilderness'.[34] That same freedom, he states, can be found in cyberspace, which has no inherent limitations and no social restraints on behaviour.[35]

A further description of cyberspace was made by John Perry Barlow, a retired Wyoming cattle rancher, former lyricist for the American rock band the Grateful Dead, and co-founder of the Electronic Frontier Foundation, an organisation which promotes freedom of expression in digital media. Since May of 1998 he has been a Fellow at Harvard Law School's Berkman Centre for Internet and Society.[36]

Barlow also compares cyberspace to the Wild West in terms of its lawlessness and freedom:

28 Dave Healy, 'Cyberspace and Place: The Internet as Middle Landscape on the Electronic Frontier', in David Porter, ed., *Internet Culture* (New York and London: Routledge, 1997), [hereafter Healy, 'Cyberspace and Place'], p. 57.
29 Healy, 'Cyberspace and Place', p. 57.
30 Healy, 'Cyberspace and Place', p. 55.
31 Turner, 'The Significance of the Frontier'.
32 Mary Midgley, *Science as Salvation: A Modern Myth and its Meaning* (London and New York: Routledge, 1992), [hereafter Midgley, *Science as Salvation*], pp. 193-194.
33 Midgley, *Science as Salvation*, p. 193-194.
34 Healy, 'Cyberspace and Place', p. 65.
35 Healy, 'Cyberspace and Place', p. 56.
36 *Electronic Frontier Foundation* website, http://homes.eff.org/~barlow/ [accessed 6 July 2009].

> Cyberspace, in its present condition, has a lot in common with the 19th Century West. It is vast, unmapped, culturally and legally ambiguous, verbally terse, hard to get around in, and up for grabs. Large institutions already claim to own the place, but most of the actual natives are solitary and independent, sometimes to the point of sociopathy. It is, of course, a perfect breeding ground for both outlaws and new ideas about liberty.[37]

If space is perceived as a frontier, then following Turner's thesis of the American identification with the frontier, exploring space becomes a form of national self-expression. Turner's argument could explain the emotional and territorial reaction described by Naughton when, in 1957, the Soviets preceded them into their final frontier. It could then be argued that if cyberspace is perceived as the new frontier, it also becomes a space to be explored, where the console cowboy is once more free from the restrictions of civilisation. Once again an attempt to invade the territory provoked a strongly emotional reaction. In 1995 the US Congress passed a law, the Communications Decency Act, attempting to regulate what could be transmitted over the Internet. In what became known as the Black World Wide Web Protest, over 20,000 websites turned their web pages black for forty-eight hours in protest, and Senators and Congressmen were inundated with angry letters and emails. There were strong legal challenges from the Electronic Frontier Foundation and the American Civil Liberties Union, which prompted Barlow to write 'A Declaration of the Independence of Cyberspace', addressed to 'Governments of the Industrial World', containing the lines,

> We have no elected government, nor are we likely to have one, so I address you with no greater authority than that with which liberty itself always speaks. I declare the global social space we are building to be naturally independent of the tyrannies you seek to impose on us.[38]

Eventually the act was declared unconstitutional by the Supreme Court in 1997.[39]

Midgley believes there is a chronic confusion between the possibility of inner, spiritual freedom and the possession of an outside, physical territory which one could, if one felt like it, always invade. She believes that overcrowding in the world could be behind this.[40] Healy also considered the hypothesis that the source of this love of freedom and the desire to go to

---

37 Barlow, 'Crime and Puzzlement'.

38 John Perry Barlow, 'A Declaration of the Independence of Cyberspace', *Electronic Frontier Foundation* Website, http://homes.eff.org/~barlow/Declaration-Final.html [accessed 6 July 2009].

39 Naughton, *A Brief History of the Future*, p. 42.

40 Midgley, *Science as Salvation*, p. 194.

the land beyond the frontier is an attempt to escape an over-crowded world. He believed this to be true of the migration to the American West and argued that the desire to physically go into outer space was another such attempt.[41]

In the 1950s Carl Gustav Jung believed that the people of the time also sought to escape from an over-crowded world via space. Writing about the thousands of Unidentified Flying Object (UFO) sightings that occurred in the 1940s, he said that in the prison of an Earth that was growing too small and threatened by the hydrogen bomb, people looked for help from extra-terrestrial sources since it could not be found on Earth, 'Hence there appears "Signs in the heavens", superior beings in the kind of space ships devised by our technological fantasy'.[42] In times of stress he believed that people looked to the heavens for salvation as they always had, 'beyond the realm of earthly organisations and powers into the heavens, into interstellar space, where the rulers of human fate, the gods, once had their abode in the planets'.[43]

The sacredness of the celestial realm has been a feature of cultures as far back as the third millennium BCE, in Mesopotamia, where, as described by Rochberg, astral phenomena were seen as manifestations of certain gods and goddesses.[44] This association of celestial bodies with the divine was incorporated into Greek culture where the *Epinomis*, written either by the Athenian philosopher Plato (428-348 BCE), or his student, Philip of Opus, was sympathetic to Mesopotamian beliefs, and supported the introduction of their astral deities to Greece.[45] He believed the planets were gods and their movements revealed divine Reason.[46] In the Myth of Er, Plato described heaven as beyond concentric spheres, of which Earth is at the centre. Each sphere was associated with a celestial body. This idea found its way into Gnosticism and became part of the medieval Christian worldview. In some Hermetic texts, heaven was beyond the eighth sphere, the sphere of the fixed stars, which was outside seven imperfect planetary spheres.[47]

41 Healy, 'Cyberspace and Place', p. 56.

42 Carl Gustav Jung, *Civilisation in Transition, The Collected Works of C.G. Jung, Volume 10*, Trans. R.F.C. Hull (London and Henley: Routledge and Kegan Paul Ltd., 1964) [hereafter Jung, *Civilisation in Transition*], p. 616.

43 Jung, *Civilisation in Transition*, p. 610.

44 Francesca Rochberg, *The Heavenly Writing: Divination, Horoscopy and Astronomy in Mesopotamian Culture* (Cambridge: Cambridge University Press, 2004), p. 46.

45 Plato, *Epinomis*, in John M. Cooper and D.S. Hutchinson, eds., *Plato: Complete Works*, trans. Richard D. McKirahan (Indiana: Hackett Publishing Company, 1997), 987A, p. 1628.

46 Plato, *Timaeus*, in John M. Cooper and D.S. Hutchinson, eds., *Plato: Complete Works*, trans. Donald J. Zeyl (Indiana: Hackett Publishing Company, 1997), 40A-41A, 47B-C.

47 'I, Poemandres, the Shepherd of Men', *The Corpus Hermeticum*, trans. G.R.S. Mead, online at http://www.gnosis.org/library/hermes1.html [accessed 15 July 2009].

Ptolemy in the second century believed that the spheres grew progressively more pure as they ascended.[48] This system was used by Dante in his *Divine Comedy*, where heaven is in the Empyrean or tenth sphere. It is beyond time, space and matter. Each of the seven inner spheres relate to a planet and become progressively more perfect as Dante moves out from earth towards heaven.[49] At the eighth sphere of the fixed stars he sees visions of Christ, the Virgin Mary and the saints. He believes that the power of the constellations is drawn from God. At the ninth sphere, the Primum Mobile, the abode of the angels, Dante sees God as a point of light surrounded by nine rings of angels.[50] His description of the ninth sphere bears some resemblance to Stenger's description of cyberspace, where we become creatures of coloured light in motion, pulsing with golden particles, and angels for eternity.[51] For Dante, God is at the Empyrean or tenth sphere. Just as with Plato, Ptolemy and in the Gnostic scriptures, God is beyond the solar system, in outer space.

In the twentieth and twenty-first centuries, the human relationship with space underwent a dramatic change as humanity attempted to physically travel into outer space or, as Noble puts it, into 'what used to be called heaven'.[52] He believes that the enchantment of space flight is 'fundamentally tied to the other-worldly concept of heavenly ascent'.[53] Midgley supports this view, pointing out that unmanned space probes have been far more useful than manned spacecraft, yet humans continue to travel into space.[54] To support Noble's argument that space is the new heaven, he gives the example of the space window in Washington Cathedral. In February 1974 a stained-glass 'space window' was officially installed at the Cathedral, containing a two-inch diameter lunar rock sample brought back on Apollo 11. The *NASA Headquarters Weekly Bulletin* announced the event, pointing out that the dean of the Cathedral 'will preach on the spiritual significance and the religious implications of the first journey from the planet Earth'.[55] Space travel, or Noble's 'heavenly ascent', is an attempt not just to travel alive into the realm of the gods, but also to return. Logging

48 Claudius Ptolemy, *Almagest*, trans. G.J. Toomer (London: Duckworth, 1984), 1.1, p. 36.

49 Dante Alighieri, *The Divine Comedy*: *Paradiso*, trans. Henry Francis Cary (London: Wordsworth Editions Ltd., 2009) [hereafter Dante, *Paradiso*], Cantos I-21.

50 Dante, *Paradiso,* Cantos 22-33.

51 Stenger, 'Mind is a Leaking Rainbow', p. 52.

52 David F. Noble, *The Religion of Technology: The Divinity of Man and the Spirit of Invention*, second edition (New York: Penguin Books, 1999 [1997]), [hereafter Noble, *The Religion of Technology*], p. 115.

53 Noble, *The Religion of Technology*, p. 115.

54 Midgley, *Science as Salvation*, p. 185.

55 'Invitation to Dedication of Space Window at Washington Cathedral', *NASA Headquarters Weekly Bulletin*, July 15 1974, cited in Noble, *The Religion of Technology*, p. 136.

into cyberspace has been seen in the same light. Speaking of cyberspace, Michael Benedikt wrote in 1991:

> another life world, a parallel universe, offering the intoxicating prospect of actually fulfilling — with a technology very nearly achieved — a dream thousands of years old: the dream of transcending the physical world, fully alive, at will, to dwell in some Beyond — and to be empowered or enlightened there, alone or with others, and to return.[56]

Travelling body and soul to the divine realm is not a new concept. In his letter to the Corinthians, St Paul wrote that at the end of days, the faithful shall enter heaven with body and soul intact, 'the dead shall be raised incorruptible'.[57] What is different about the attempt to do so via technology in Benedikt's opinion is that now it is possible to travel there alive and to return. But there is also a difference between the heaven reached through the use of a spacecraft and that reached through the use of a computer. While the journey to space involved leaving Earth, with cyberspace the person enters heaven without leaving their chair. What is being promised in the Scriptures is immortality, and if once that involved the sky, cyberspace makes it possible on Earth.

The achievement of true immortality, which implies the winning of a battle over this world of death and decay, including the physical body, is the preoccupation of Transhumanism. Transhumanism, or T+, is an international intellectual and cultural movement supporting the use of science and technology to improve human mental and physical characteristics and capacities, and to eliminate disease and suffering. According to the Transhumanist website, *Humanity+*, the core Transhumanist activity value is exploring the realm of the posthuman:

> Transhumanists view human nature as a work-in-progress, a half-baked beginning that we can learn to remould in desirable ways. Current humanity need not be the endpoint of evolution. Transhumanists hope that by responsible use of science, technology, and other rational means we shall eventually manage to become posthuman, beings with vastly greater capacities than present human beings have.[58]

As a way of prolonging human life indefinitely, the transhumanists propose uploading human minds to computers, where individuals can live forever in cyberspace, or even in the real world, by controlling a robotic version of themselves from their uploaded mind. Hans Moravec, a faculty member of the Robotics Institute at Carnegie Mellon University, describes how the

---

56 Michael Benedikt, ed., *Cyberspace: First Steps* (Massachusetts: MIT, 1991), p. 131.
57 1 Corinthians 15:52, the Bible (King James Version).
58 Anon., 'Transhumanist Values: What is Transhumanism?' http://humanityplus.org/learn/philosophy/transhumanist-values [accessed 1 July 09].

individual moves into cyberspace: The brain is connected by cables to a computer. As the brain withers with age, the computer assumes its functions. In time the brain dies and the person goes on living in the computer. As long as enough back-up copies are made, death is unlikely.[59]

One transhumanist group, the Extropy Institute, was founded in 1990 by Tom Bell and Max More, followed by similar organisations in different parts of the world, such as Transvision in Europe and Aleph in Sweden.[60] Extropianism is the belief that science and technology will someday allow people to live indefinitely and that humans alive today have a good chance of seeing that day. The Extropy Institute describes itself as 'the original force behind the philosophy and global cultural movement of Transhumanism'.[61] In 2006, the board of directors of the Extropy Institute made a decision to close the organisation, stating that its mission was essentially complete, in that they had developed a culture for activists, energised and devoted to bringing these ideas to the public, and created a website that would continue to provide a place for Transhumanists to meet for 'challenging and creative discussions about the future'.[62]

The aims of Transhumanism and Moravec sound strangely like the promise of the book of Revelation, 'And God shall wipe away all tears from their eyes; and there shall be no more death, neither sorrow nor crying, neither shall there be any more pain: for the former things are passed away'.[63] Davis, writing about Moravec, says, 'Here again is the real wonder: that information technology allows even the most hard-core materialists to ruminate once again on the ancient dream of slipping the incorporeal spark of the self through the jaws of death unscathed'.[64] Just as with Stenger, cyberspace is being perceived as an earthly heaven where the digital faithful can live forever.

Other writers have taken the relationship between cyberspace and the divine a step further. Two further ideas have crept into cyberspace theory. The first of these is that cyberspace has become God, the second that it enables human beings to collectively become God. The idea of cyberspace as God has been posited by Ward who called it 'The ultimate secularisation of the divine',[65] and Zaleski, who asked, 'Will the WWW evolve into God?'[66] Naughton points out that since the Internet was designed so that it could

---

59 Hans Moravec, *Mind Children: The Future of Robot and Human Intelligence* (Cambridge, Mass: Harvard University Press, 1988), pp. 121-124.

60 See their websites at http://www.transvision2007.com/ and http://www.aleph.se/Trans/ [accessed 09 July 2010].

61 Anon., http://www.extropy.org/ [accessed 13 July 2009].

62 Anon., http://www.extropy.org/ [accessed 13 July 2009].

63 Revelation 21:4, the Bible (King James Version).

64 Davis, *TechGnosis*, p. 148.

65 Graham Ward, *The Postmodern God* (Oxford: Blackwell Publishers, 1997), p. xvi.

66 Jeff Zaleski, *The Soul of Cyberspace: How New Technology is Changing Our Spiritual Lives* (New York: HarperCollins, 1997), p. 20.

continue to pass messages even if large chunks of it were irretrievably damaged, it is no longer under the control of humanity. He concludes that since it is resilient enough to withstand a nuclear attack, it is unlikely that it could be switched off.[67] In this view of cyberspace, it has an existence of its own outside human control.

The concept of artificial intelligence surpassing human intelligence is the basis of much of the writing of Ray Kurzweil. He proposes the physical merging of people and machines in his 2005 book, *The Singularity is Near: When Humans Transcend Biology*.[68] The Singularity is the point in the future at which machine intelligence outstrips human brainpower. It is estimated by Kurzweil that this will happen by the year 2030.[69] The term 'Singularity' was first defined by Vernor Vinge and has been developed by Kurzweil, a computer engineer with an impressive track record in invention.[70] In this future era, Strong Artificial Intelligences (SAIs) and cybernetically augmented humans (cyborgs) will replace the current biological model. Once again it is proposed that the machine-enhanced human can live forever and always experience health and happiness.[71] The SAIs and cyborgs will overwhelm the universe and saturate its matter with intelligence. Kurzweil predicts that complete universal saturation can occur by the end of the twenty-second century.[72] In a debate at the Massachusetts Institute of Technology (MIT) on 30 November 2006, Kurzweil and David Gelernter, professor of computer science at Yale, held a debate on the topic, 'Will Machines Become Conscious?' The debate also asked if machines could develop spirituality. There was no winner of the debate, but Gelernter, who argued against the motion, based his arguments on the belief that the human brain can not be reduced to a machine, a necessary criteria for the fulfilment of Kurzweil's visions.[73]

In attempting to achieve immortality and thus victory over life and death, humanity is not just emulating God or the gods; they are seeking to become God. This aspiration is put into words by Artificial Intelligence (AI) guru Earl Cox. Cox claims that because of technological achievements, humans have evolved beyond their original design and outdistanced their creator, and that the uploaded minds stored in cyberspace will collectively

67 Naughton, *A Brief History of the Future*, p. 36.
68 Ray Kurzweil, *The Singularity is Near: When Humans Transcend Biology* (New York: Viking Press Inc, 2005), [hereafter Kurzweil, *The Singularity is Near*].
69 Kurzweil, 'The Singularity', *KurzweilA1.net*, at http://www.kurzweilai.net/meme/frame.html?main=memelist.html?m= 1%23696 [accessed 09 July 2010].
70 Anon., http://www.kurzweilai.net/ and http://www.kurzweiltech.com/kti flash.html [accessed 13 July 2009].
71 Kurzweil, *The Singularity is Near*, pp. 24-30.
72 Kurzweil, *The Singularity is Near*, pp. 353-366.
73 Rodney Brooks, David Gelernter and Ray Kurzweil, 'Will Machines Become Conscious', http://www.kurzweilai.net/meme/frame.html?main=memelist.html?m=4%23688 [accessed 13 July 2009].

become more powerful than current concepts of the divine. In his words, 'Such a combined system of minds, representing the ultimate triumph of science and technology, will transcend the timid concepts of deity and divinity, held by today's theologians'.[74] The idea that mankind could become god or god-like was considered (in a less extreme version) by Michael Heim in 1993, 'What better way to emulate God's knowledge, than to generate a virtual world constituted by bits of information? Over such a cyber world human beings could enjoy a god-like instant access'.[75] Heim is implying not that cyberspace is God, but that by giving people the power to enter this digital heaven and return at will, it empowers humans to become God.

From the evidence in his writings, William Gibson appears to have been aware of this potential development in the perception of cyberspace. In *Mona Lisa Overdrive*, the third and final novel in the series that began with *Neuromancer*, an artificial intelligence (a computer), when asked if the Matrix (cyberspace) is God, stated that it would be more accurate to say that the Matrix has a god since this god's omniscience and omnipotence are assumed to be limited to the Matrix. The being is not assumed to be immortal, as its existence is dependent on the life of the Matrix. 'Cyberspace exists, in so far as it can be said to exist, by virtue of human agency'.[76]

So in the view of its creator, cyberspace, like heaven, like the celestial realms, contains a divine presence. One feature of this divine presence is that arguably God or the gods are being perceived as confined, with their power limited and confined by physical technology Building on Gibson's argument. In this scenario, what appears to have taken place is a type of Genesis in reverse, where man has created God, a suggestion that has led theorists like Cox and Kurzweil to the view that humanity has achieved the traditionally god-associated characteristics of omnipotence and immortality.

Bibliography

Aggelou, George, *Mobile Ad-hoc Networks* (New York: McGraw-Hill, 2005).

Anon., 'Definition of Cloud', *PC Magazine Encyclopedia*, http://www.pcmag.com/encyclopedia_term/0,2542,t=cloud&i=39847,00.asp [accessed 7 July 2009].

Anon., 'Definition of Cloud Computing', *PC Magazine Encyclopedia*, http://www.pcmag.com/encyclopedia_term/0,2542,t=cloud+computing&i=57964,00.asp [accessed 7 July 2009].

---

74 Earl Cox and Gregory Paul, *Beyond Humanity: CyberEvolution and Future Mind* (Cambridge: Charles River Media, 1996), pp. 1-7.

75 Michael Heim, 'The Erotic Ontology of Cyberspace' in Michael Benedikt, ed., *Cyberspace: First Steps* (Cambridge, Mass: MIT Press, 1992 [1991]), pp. 59-80, 69.

76 William Gibson, *Mona Lisa Overdrive* (London: Voyager, 1995 [1988]), p. 107.

Anon., 'From Arpanet to World Wide Web', Internet History, University of Leiden, http://www.leidenuniv.nl/letteren/ internethistory/index.php3c=5.htm [accessed 4 July 2009].
Anon., *World Internet Statistics*, http://www.internetworldstats.com/stats.htm [accessed 5 July 2009].
Anon., 'Transhumanist Values: What is transhumanism?', http://humanityplus.org/learn/philosophy/transhumanist-values [accessed 1 July 2001].
Anon., *Extropy Institute*, http://www.extropy.org/ [accessed 13 July 2009].
Barlow, John Perry, 'Crime and Puzzlement', http://w2.eff.org/Misc/Publications/John_Perry_Barlow/HTML/crime_and_puzzlement_1.html [accessed 5 July 2009].
Barlow, John Perry, 'A Declaration of the Independence of Cyberspace', *Electronic Frontier Foundation Website*, http://homes.eff.org/~barlow/Declaration-Final.html [accessed 6 July 2009).
Bell, David, 'Cybercultures Reader: A User's Guide', in David Bell and Barbara M. Kennedy, eds., *The Cybercultures Reader* (London: Routledge, 2000), pp. 1-12.
Benedikt, Michael, 'Cyberspace: Some Proposals', in Michael Benedikt, ed., *Cyberspace: First Steps* (Massachusetts: MIT, 1991), pp. 119-224.
Bergman, Michael K., 'The Deep Web: Surfacing Hidden Value', *The Journal of Electronic Publishing* (August 2001), Vol. 7(1), http://quod.lib.umich.edu/cgi/t/text/text-idx?c=jep;view=text;rgn=main;idno=3336451.0007.104 [accessed 7 July 2009].
Brooks, Rodney, David Gelernter and Ray Kurzweil, 'Will Machines Become Conscious', http://www.kurzweilai.net/meme/frame.html?main=memelist.html?m=4%23688 [accessed 13 July 2009].
The Corpus Hermeticum, 'I, Poemandres, the Shepherd of Men', trans. G. R. S. Mead, http://www.gnosis.org/library/hermes1.html [accessed 15 July 2009].
Cox, Earl and Gregory Paul, *Beyond Humanity: CyberEvolution and Future Mind* (Cambridge: Charles River Media, 1996).
Davis, Erik, *TechGnosis: Myth, Magic and Mysticism in the Age of Information* (London: Serpents Tail, 2004 [1998]).
Gibson, William, *Neuromancer* (London: Voyager, 1995 [1984]).
Gibson, William, *Count Zero* (London: Voyager, 1995 [1986]).
Gibson, William, *Mona Lisa Overdrive* (London: Voyager, 1995 [1988]).
Healy, Dave, 'Cyberspace and Place: The Internet as Middle Landscape on the Electronic Frontier', in David Porter, ed., *Internet Culture* (New York and London: Routledge, 1997), pp. 55-68.
Heim, Michael, 'The Erotic Ontology of Cyberspace' in Michael Benedikt, ed., *Cyberspace: First Steps* (Cambridge, Mass: MIT Press, 1992 [1991]), pp. 59-80, 69.
'Invitation to Dedication of Space Window at Washington Cathedral', *NASA Headquarters Weekly Bulletin*, July 15 1974, http://articles.adsabs.harvard.edu//full/1977NASSP4019.....B/0000118.000.html [accessed 12 July 2009].
Jung, Carl Gustav, The *Collected Works of C. G. Jung, Vol. 10, Civilisation in Transition*, trans. R. F. C. Hull (London and Henley: Routledge and Kegan Paul Ltd., 1964).
Karaflogka, Anastasia, *E-religion: A Critical Appraisal of Religious Discourse on the World Wide Web* (Equinox: London, 2006).

Ray Kurzweil, *The Singularity is Near: When Humans Transcend Biology* (New York: Viking Press Inc, 2005).
Leiner, Barry M, G. Vinton, David D. Cert, Robert E. Clarke, Leonard Kleinrock Kahn, Daniel C. Lynch, Jon Postel, Larry G. Roberts, Stephen Wolff, 'A Brief History of the Internet', *The Internet Society*, http://www.isoc.org/internet/history/brief. shtml [accessed 4 July 2009].
Midgley, Mary, *Science as Salvation: A Modern Myth and its Meaning* (London and New York: Routledge, 1992).
Moravec, Hans, *Mind Children: The Future of Robot and Human Intelligence* (Cambridge, Mass: Harvard University Press, 1988).
Naughton, John, *A Brief History of the Future: The Origins of the Internet*, second edition (London: Phoenix, 2000 [Weidenfeld & Nicolson, 1999]).
Noble, David F., *The Religion of Technology: The Divinity of Man and the Spirit of Invention*, second edition (New York: Penguin Books, 1999 [1997]).
Plain, Charlie, 'A Dim Light Shines on Dark Matter', *NASA* website, http://www.nasa.gov/missions/deepspace/chandra_dark_matter_halo.htm l [accessed 7 July 2009].
Plato, *Epinomis*, in John M. Cooper and D.S. Hutchinson, eds., *Plato: Complete Works,* trans. Richard D. McKirahan (Indiana: Hackett Publishing Company, 1997).
Plato, *Timaeus,* in John M. Cooper and D.S. Hutchinson, eds., *Plato: Complete Works,* trans. Donald J. Zeyl (Indiana: Hackett Publishing Company, 1997).
Plato, *Republic*, in John M. Cooper and D.S. Hutchinson, eds., *Plato: Complete Works*, translated by G.M.A. Grube and C.D.C. Reeve (Indiana: Hackett Publishing Company, 1997).
Ptolemy, Claudius, *Almagest*, trans. G.J. Toomer (London: Duckworth, 1984).
Rochberg, Francesca, *The Heavenly Writing: Divination, Horoscopy and Astronomy in Mesopotamian Culture* (Cambridge: Cambridge University Press, 2004).
Stenger, Nicole, 'Mind is a Leaking Rainbow', in Michael Benedikt, ed., *Cyberspace: First Steps* (Massachusetts: MIT 1991), pp. 49-58.
Turner, Frederick Jackson, 'The Significance of the Frontier in American History', *Proceedings of the State Historical Society of Wisconsin, December 14, 1893,* Published online in Frederick Jackson Turner, *The Frontier in American History*, http://xroads.virginia.edu/~HYPER/TURNER/ [accessed 5 July 2009].
Ward, Graham, *The Postmodern God* (Oxford: Blackwell Publishers, 1997).
Wertheim, Margaret, *The Pearly Gates of Cyberspace* (London: W.W. Norton & Co. Ltd., 1999).
Zaleski, Jeff, *The Soul of Cyberspace: How New Technology is Changing Our Spiritual Lives* (New York: HarperCollins, 1997).

Contributors

Pauline Bambrey is currently a PhD candidate in Social Anthropology at the University of Wales, Trinity Saint David, conducting research into The Beltane Fire Society, Edinburgh. Her research looks at how the use of the body in ritualised performance establishes identity within a community.

Glenford Bishop is a fellow of the Royal Astronomical Society. He is currently a lecturer at the University of Plymouth and a PhD student with the Sophia Centre, University of Wales, Trinity Saint David. Glenford has a clinical and educational background in occupational therapy and dramatherapy, focusing on psychodynamic approaches. These latter approaches have informed a literary method for re-appraising the Mummers Play and the mystery of its origins.

Nicholas Campion is Senior Lecturer in the School of Archaeology, History and Anthropology and Director of the Sophia Centre for the Study of Cosmology in Culture at the University of Wales, Trinity Saint David. He is course director of the University's MA in Cultural Astronomy and Astrology. His publications include the two-volume *History of Western Astrology* (Continuum, 2009).

Frances Clynes was awarded the MA in Cultural Astronomy and Astrology from Bath Spa University in 2007, and is currently working on a PhD at the University of Wales, Trinity Saint David, where she is also a part-time tutor. She has a BSc and MSc from Trinity College Dublin and works as a computing lecturer in a Dublin college.

Martin Gansten is a historian of religion specialising in astrological and other divinatory traditions in India as well as the west. He has taken particular interest in the South Indian phenomenon of *nāḍī* reading, in the development of the method of primary directions (*aphesis, at-tasyīr*) from the early Greek period up to the present day, and in the mutual influences of Perso-Arabic and Indian astrological traditions in the Middle Ages. He is based in Lund University, Sweden.

Ronald Hutton is Professor of History at Bristol University, and the author of thirteen books on different aspects of British and overseas history. Those most relevant to this collection are *The Rise and Fall of Merry England* (Oxford University Press, 1994) and *The Stations of the Sun: A History of the Ritual Year in Britain* (Oxford University Press, 1996).

Helen R. Jacobus is a PhD candidate at the University of Manchester researching the Dead Sea Scroll that is the subject of this paper and calendars at Qumran. Her publications include 'The Date of Purim and Calendars in the Book of Esther', in Jonas Vaiškūnas, ed., *Archaeologica Baltica* (Klaipėda University Press, 2008, pp. 114-118), and '4Q318: A Jewish Zodiac Calendar at Qumran', in Charlotte Hempel, ed., *The Dead Sea Scrolls: Texts and Contexts, Studies on the Texts of the Desert of Judah, Volume 90* (Brill, forthcoming, 2010).

Jane Ridder-Patrick has a MSc in history from the University of Edinburgh, where she is currently pursuing a PhD in Astrology in Early Modern Scotland. Trained in pharmacy, naturopathy, herbal medicine and psychotherapy, she is author of *A Handbook of Medical Astrology*.

Lionel Sims is Head of Anthropology at the University of East London, Vice-President of the Society of European Archaeoastronomy (SEAC), and is a member of the Stonehenge Round Table hosted by English Heritage.

Mark Williams is a research fellow at Peterhouse, Cambridge, where he teaches Latin and Medieval Irish and Welsh language and literature. His first book, *Fiery Shapes: Celestial Portents and Astrology in Ireland and Wales, 700-1700*, is forthcoming from Oxford University Press. He is currently writing a cultural history of the gods of Irish mythology.

The Sophia Centre

The Sophia Centre was set up with funding from the Sophia Trust and is located within the University of Wales, Trinity St David's School of Archaeology, History and Anthropology. It has a wide-ranging remit to investigate the role of cosmological, astrological and astronomical beliefs, models and ideas in human culture, including the theory and practice of myth, magic, divination, religion, spirituality, politics and the arts. The Centre teaches the MA in Cultural Astronomy and Astrology via distance learning online, and also supervises MPhil and PhD research. There is no need to live in the UK to study at the Sophia Centre.

Much of the Centre's work is historical, but it is equally concerned with contemporary culture and lived experience. If you are interested in the way the sky is used to create meaning and significance, then the Sophia Centre may be the best place for you to study. By joining the Sophia Centre, you enter a community of like-minded scholars whose aim is to explore humanity's relationship with the cosmos.

For further information about the Sophia Centre see the website at www.lamp.ac.uk/sophia, or contact Nick Campion, the Course Director, at the details below.

The Sophia Centre
Department of Archaeology and Anthropology
University of Wales, Trinity Saint David
Ceredigion
Wales SA48 7ED
United Kingdom
Email: n.campion@lamp.ac.uk

Index